Customer-Centric Design

This book presents a cutting-edge customer-centric design approach, equipping readers with specific tools to effectively analyze customer needs and develop top-tier designs while taking pricing and competition into account. The meticulously chosen tools, rigorously tested by the authors, are derived from Quality Function Deployment (QFD), a proven method with a 50-year track record of successful implementation across various industry sectors.

Customer-Centric Design: Based on QFD Principles introduces a contemporary guide to the design principles of Blitz QFD®, a groundbreaking methodology developed by the QFD Institute over 25 years back. This book is crafted to optimize customer processes, leading to heightened success and increased business opportunities. By tackling customer concerns and concentrating on top-line revenue growth through the sale of high-value goods and services, the book offers a strategic approach to business development. Additionally, it focuses on the fundamentals of QFD prioritization, including the analytic hierarchy process, enabling more precise measurement of customer priorities and critical design decisions. Moreover, the book is compliant with the new ISO 16355 for QFD standard, ensuring that it aligns with the latest industry requirements. Relevant references will also be provided for further exploration.

Product managers, engineers, and technologists will find this book particularly valuable, as it offers user-friendly methods and tools for validating marketing requirements and conducting market studies independently, as well as strategies to efficiently use these tools within tight time constraints.

Best on Quality: Advancing Quality for Humanity
Series Editors: Elizabeth A. Cudney, Grace Brannan and Hiroe Tsubaki

Each volume in the series is a collection of quality articles written by IAQ members, their collaborators, and invited experts. The books will explore quality concepts, practices, successes, and challenges facing the 21st century and beyond.

Each book will focus on bringing quality to humanity and expand from quality applications in the manufacturing and service industries to emphasizing and creating higher value for people and their quality of life with lesser use of resources. But more so, the books will talk about humanity as the main stakeholder for quality. Beyond satisfying customers, quality initiatives must fulfill the needs of society, their country, and the planet and be sustainable.

This new series, which will be a partnership between IAQ and CRC Press/Taylor and Francis, will include research, case studies, practical applications, and concepts on a variety of subject areas all related to quality for humanity.

If you are interested in writing or editing a book for the series or would like more information, please contact Cindy Carelli, cindy.carelli@taylorandfrancis.com.

Customer-Centric Design
Based on QFD Principles
David Menichelli and Glenn H. Mazur

For more information on this series, please visit: www.routledge.com/Best-on-Quality-Advancing-Quality-for-Humanity/book-series/IAQBOQ

Customer-Centric Design

Based on QFD Principles

David Menichelli and Glenn H. Mazur

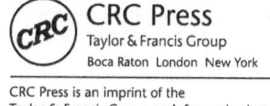

CRC Press
Taylor & Francis Group
Boca Raton London New York

CRC Press is an imprint of the
Taylor & Francis Group, an **informa** business

First edition published 2025
by CRC Press
2385 NW Executive Center Drive, Suite 320, Boca Raton FL 33431

and by CRC Press
4 Park Square, Milton Park, Abingdon, Oxon, OX14 4RN

CRC Press is an imprint of Taylor & Francis Group, LLC

ISBN: 978-1-032-89840-7 (hbk)
ISBN: 978-1-032-89754-7 (pbk)
ISBN: 978-1-003-54484-5 (ebk)

DOI: 10.1201/9781003544845

Typeset in Times
by Newgen Publishing UK

Contents

Foreword

Customer First! So often said, not enough done!

In today's competitive landscape, teams that do not invest enough time in truly listening to the voice of the customer often fail to address the real demands of the market. It is not enough to develop a product that is technically excellent – better is not always synonymous with success. What matters most is whether the product effectively responds to the real needs of the customer. Sometimes a product that may not be the most technically advanced, but that meets the customer's needs, will far outperform in terms of revenue compared to a superior technical solution that lacks this alignment. Truly listening to the voice of the customer requires a strategic approach for integrating the voice of the customer at specific stages of product development. Too often, product development teams fall in love with a particular technical solution – such as a new user interface or a feature that improves the accuracy of a measurement – that may seem impressive at first. However, these additions, if not properly discussed with customers, often increase the final price of the product and, once implemented, can be met with indifference or disappointment, resulting in commercial failure.

This book, *Customer-Centric Design Based on QFD Principles*, provides a comprehensive framework that embodies a structured, customer-first approach. It provides the tools necessary to not only capture customer needs and wants, but also to prioritize those needs in a way that drives innovation and minimizes wasted effort. Its structure is deeply rooted in the Quality Function Deployment (QFD) methodology, a proven technique that links product features directly to customer needs.

As someone who has spent nearly three decades at the intersection of high-tech innovation and product development, I understand the challenges of designing and manufacturing innovative products under time pressure. My career, which includes leadership roles in international companies, including my current position as Chief Technology Officer at Evatec AG in Switzerland, has taught me the importance of not only having a vision for technology, but also aligning that vision with customer-driven goals.

I first met one of the authors – Dr. Menichelli – in the early stages of my career at the National Institute of Nuclear Physics in Italy, where we were both involved in the research and development of radiation detectors for high-energy physics applications. This period marked the beginning of our professional connection. However, our paths eventually diverged when we both decided to move to industry: I moved to Switzerland and joined Oerlikon Solar, a company that developed and marketed turnkey solutions for the production of thin-film solar panels, while he moved to Germany and joined IBA Dosimetry. The time I spent at the Institute of Nuclear Physics and working with David taught me methods and methodologies of research and development, but my real education in customer interaction began, of course, when I left academia and started working in product development in industry. It was at Oerlikon Solar, and then at Alstom, General Electric, ABB, and now at Evatec, that I learned the value of aligning the developed product with customer needs. It really is a continuous learning experience.

Despite the first impression that this book may seem like a theoretical or statistics-based product development manual, the principles outlined in this book are not theoretical exercises, but practical tools that can be implemented in real-world product development scenarios. While the concepts discussed here may not be entirely new, they are presented in a systematic and logical way that makes them highly accessible and actionable. In addition, the book offers a dual approach – both qualitative and more "scientific" – that accommodates readers from different backgrounds and allows them to extract and apply the aspects that best suit their needs. For those seeking a deeper understanding, the book also provides references for further reading, offering a path for the most demanding readers to explore these ideas in more detail. This thoughtful structure ensures that everyone, whether seeking a practical guide or a deeper analytical framework, can benefit from the insights provided.

One of the central themes of this book is the idea of the "minimum merchandisable product". In a world where time is often the most valuable resource, developing a product that addresses the top needs of the most critical customer segments is a key to success. As this book explains, understanding customer needs and translating them into product features is essential, but equally important is the ability to prioritize and focus on delivering those features efficiently. Overengineering, while tempting, can be detrimental, especially in industries where the window for market entry is narrow.

The core principle emphasized in this book is the importance of listening to the voice of the customer. This approach is more than just gathering superficial feedback – it requires a deep understanding of the customer's pain points, challenges, and needs. The methods outlined in the book provide a set of tools that enable professionals to systematically capture this voice and ensure that customer feedback becomes a driving force in product development.

The authors present several approaches to listening to the voice of the customer. On the one hand, a more rigorous, structured method can be implemented, involving detailed data collection through interviews, focus groups, or even on-site observations (*gemba* visits). This method allows for an accurate and thorough understanding of customer needs and ensures that no critical detail is overlooked. For companies willing to invest more time and effort, this approach provides solid, quantifiable insights that directly shape product design and decision-making.

In many cases, however, a more qualitative and flexible approach is the only viable option, although it carries certain risks. This may involve faster, less formal interactions with customers, gathering their feedback through interviews, surveys, or informal discussions. While this approach may lack the precision of more structured methods, it can still provide valuable insights if used thoughtfully. As described in the following chapters, it's important to use these techniques carefully to ensure that real customer needs are captured accurately and without bias, even when working under tight time constraints. It is always imperative to go back to the customer and test the assumptions.

The authors also provide a guide for transitioning product development from understanding customer needs to crafting a product concept, where the focus shifts to translating those insights into a concrete solution that addresses the most critical needs. After gathering and analyzing customer feedback through visits and interviews, the unsolved needs are prioritized and used as the basis for creating a high-level product concept. This concept not only outlines what the product must do to meet the customer's desires, but also considers the constraints the design must face in order to meet the customer's expectations. Creativity is key in this phase as teams work to develop solutions that balance practicality with innovation, typically focusing on no more than three top customer needs. Functional and nonfunctional requirements guide this process: functional requirements define what the product must do to satisfy the customer's needs (desires), while nonfunctional requirements define the final performance of the product given the constraints. Tools such as tables and matrices are used to trace these requirements back to the original customer needs as prioritized after listening to the voice of the customer, helping to avoid overengineering and ensuring that the design remains focused on solving the right problems efficiently. In addition, competitive analysis plays a critical role in ensuring that the proposed solution not only meets the customer's needs, but also offers a compelling value proposition in the marketplace, balancing performance and cost.

The proof-of-concept phase, as explained in the book, is critical to validating whether a product design effectively meets customer needs and has market potential. Once the initial product concept is developed, it is presented to key customers to assess whether it addresses their challenges and whether they would be willing to pay for it. The authors advocate using a structured "pain-claim-gain" approach that highlights customer problems, presents the benefits of the solution, and emphasizes the product's unique value. Customer feedback from this phase helps refine the design to better meet expectations. After adjustments are made, the concept is presented to internal stakeholders. The authors emphasize the importance of demonstrating how the product aligns with both customer and company goals. By integrating customer insights and ensuring strategic alignment, the proof-of-concept phase confirms that the product is ready for further development and business approval.

Finally, the authors extend the principles of product design to the entire product lifecycle, emphasizing their relevance beyond the initial stages of development. These fundamental concepts are rooted in QFD.

Fostering open communication within teams is essential to success, as is the use of systematic processes such as the proposed "catch-ball" method, which encourages collaborative iteration among team members. Fair judgment and structured decision-making are emphasized to avoid cognitive biases, improve innovation, and ensure that products are aligned with market needs.

Ultimately, the methods presented are not fail-proof formulas, but practical tools that, when used with critical thinking, can significantly improve product design and development and help organizations better meet customer needs.

This book not only serves as a guide for professionals looking to deepen their understanding of customer-centric methodologies, but also encourages a cultural shift within organizations. As someone who has developed technology roadmaps and implemented innovation governance at the highest levels, I can attest to the fact that successful innovation is not a one-time effort but a sustained commitment to aligning product development with customer expectations. This requires cross-functional collaboration, continuous feedback loops, and a deep understanding of market dynamics – all of which are emphasized in the chapters that follow.

The blend of practical tools and theoretical insights presented here will be invaluable for any organization looking to enhance its product development processes. Whether you are a seasoned professional or new to the field, the QFD principles in this book will provide you with the necessary framework to create products that not only meet but also exceed customer expectations.

In closing, I am proud to support this work and encourage every reader to embrace the methodologies outlined within. As the pace of technological advancement accelerates, those who succeed will be the ones who place the customer at the heart of their innovation process.

With the best of success!

Dr. Carlo Tosi
Chief Technology Officer
Evatec AG
Switzerland

Preface

"Why can't I find a brilliant book about customer-centric product design?" I would have loved to see just one, possibly based on today's Quality Function Deployment (QFD), a method I was acquainted with. Most business methods have up-to-date, easy-to-read, and often entertaining books, so how was it possible that QFD, a methodology with 50 years of success, was still saddled with complex and paralyzingly analytic treatises which are time-consuming to read and propose methods which require, in many cases, a reorganization of the company?

This question brought me to an additional reflection. True, custom-tailoring the QFD method to a company's organization with the help of a business consultant helps a lot. But what if such a change is not yet possible? Could even a single person, working in a technology-driven environment, benefit from customer-centric methods? I firmly believe that "yes", even one effort to truly understand customers creates a pull-through benefit that seeds future improvements. My conclusion was that a book was needed to guide professionals who want to get acquainted with these powerful customer-centric methods.

I was thus inspired to capture my experience in a book. However, this experience was limited to my specific company and industry. In order to remove such idiosyncrasies and to generalize the publication for a broader audience I decided to ask Glenn H. Mazur, who had helped my company adopt QFD in the early 2010s and thereafter guided me to achieve the QFD Institute's QFD Black Belt®, to become my co-author. It was with great pleasure he accepted my proposal, and so the creation of this book started.

David Menichelli
September 2024

Acknowledgments

The authors are grateful to Sebastian Hähnel, Carlo Tosi, and Kim Stansfield for their support in reviewing the manuscript and for the helpful discussions about its content. David wishes to thank IBA Dosimetry for providing the opportunity to learn, practice and develop the QFD methodology.

About the Authors

David Menichelli received his PHD from the University of Florence in 2002, where he worked as researcher in the field of applied physics and adjunct professor of physics till 2009. Since 2010 he has been with IBA Dosimetry, where he covered several roles in R&D, product management and program management. In 2015 he received the Quality Function Deployment (QFD) Black Belt® from QFD Institute and is an expert resource for the development of ISO 16355-9. Since then, he investigates and puts into practice methods to design products based on customer-centric principles. He is the author of more than 70 peer-reviewed scientific papers and inventor of six international patents.

Glenn H. Mazur has been active in QFD since its inception in North America and has worked extensively with the founders of QFD on their teaching and consulting visits from Japan. He is a leader in the application of QFD as well as conducting advanced QFD research and is the chair for the International Symposium on Quality Function Deployment. Glenn is executive director of the QFD Institute and International Council for QFD and retired adjunct lecturer on TQM at the University of Michigan College of Engineering and retired senior member of the American Society for Quality (ASQ) and the Japanese Society for Quality Control (JSQC). He is a certified QFD Red Belt® (highest level), one of two in North America. He is a certified QFD-Architekt #A21907 by QFD Institut Deutschland. He is honorary president of the Hong Kong QFD Association and Asia QFD Association. He is convenor of the ISO Working Group which has written the ISO 16355 series standards for QFD, member of TC69/SC7 writing the ISO 20575 standard for Design for Six Sigma, and liaison to TC 312 writing the ISO/TS 23686 standard for Service Excellence and TC 207 for Environmental Management. He is an academician of the International Academy for Quality. He is the recipient of the 1998 Akao Prize® for Excellence.

1 Introduction

1.1 WHY CUSTOMER-CENTRIC PRODUCT DESIGN?

"Customer-centric design" is an approach to the design of products in which the satisfaction of customer needs plays a crucial role. Its fundamental assumption is that to systematically develop successful products, a good understanding of customers' situation is needed. This ensures that their pains are addressed by the product, their needs are solved, and they see a true value in it.

The same statement can be formulated in a negative way: systematic success cannot be based on the hope that a product will sell, or on the illusion that customers will buy what the company thinks they will like.

1.1.1 DESIGN OF INNOVATIVE PRODUCTS

The importance of customer-centric design is straightforward in the case of "innovative" products, which aim to create a competitive advantage for the company. In this case, the design process starts with an investigation of unaddressed pains, unfulfilled opportunities, or inadequate self-fulfillment in a customer's life. The drafting of product functions and features enters the design process at a later stage only, as a solution to these unaddressed issues.

Customer-centric design is essential today for the development of innovative products because, due to globalization and digital communication:

a. Technologies are increasingly available, thus reducing the entrance price for competitors.
b. Customers' expectations grow in view of global offers, which allows them to compare not only equivalent products, but products developed for different purposes too . A striking example is provided by mobile phone apps setting a high standard for the intuitiveness and usability of any software.
c. Competitors will quickly become aware of your failures and learn from them.

In these conditions the development of a "wrong product", which is badly received by the market, could create a fatal gap against competition. It is thus necessary to maximize the likelihood that the product is "right at the first attempt". This likelihood

DOI: 10.1201/9781003544845-1

cannot be 100%, since innovation always brings risks, and sometime a product idea will not be successful. However, focusing on customer needs allows us to maximize this likelihood.

Facts (a) and (b) above are responsible for another important constraint in product design: time pressure. The needs of customers and their priorities evolve quickly, as new products are constantly released on the market. As a result, the time available for product development in general, and for product design in particular, shortens. This means that it is not enough to design a product which satisfies customer needs; it must be done quickly too. For instance, in small-medium enterprises a sound product concept based on customer needs could be expected in a few months. For this reason, one goal of this book is to provide not only tools and methods for customer-centric design, but also indications in how to use them efficiently.

1.1.2 DESIGN OF PRODUCT IMPROVEMENTS

A product does not need to be innovative to be profitable, and there are non-innovative strategies that can be useful for a company – for instance, replicating successful ideas of competitors; learning from errors of a company that failed trying to create an innovative product and offering either better performances or lower price; or "tuning" an existing product, for example, to improve its market position or to sell it in a new market.

Even in these cases, it would be very dangerous to focus entirely on the product, since it is not always obvious which performance shall be improved and how much, or what decrease in price would make a difference for the customers. The answers to these questions must come from a good understanding of the customer's situation. We can mention as an example the case of a company manufacturing precision measurement instruments which wanted to further improve a high-end product suffering from strong competition. The initial idea was to make it even more accurate. However, an investigation with international customers showed that a further improvement in accuracy would not have created any additional value for them. On the other hand, they would have rather had better usability, since all products available on the market were accurate enough, but too cumbersome and inefficient to use.

1.2 RELATIONSHIP BETWEEN THIS BOOK AND QFD

Quality function deployment (QFD) is a method to ensure that the product characteristics necessary to generate customer value are included in the product design [1]. Originally, the name "Quality Function Deployment" was intended to mean "deployment of quality through deployment of quality functions", where the "quality functions" are all the company activities (like design, marketing, service, and so forth) responsible to create product quality [2]. In other words, quality is not the result of the efforts of a single team, but is achieved as an interdepartmental activity, which stretches over many departments: all "functions" must work together with the aim to create quality. Later the term "function" was used to indicate the product functions (i.e. what the product does) creating customer value (i.e. quality) as well [3].

Because of the original meaning of its name, QFD is often understood as a method that requires a reshaping and optimization of the way the company activities are carried out. This understanding represents both a great potential of QFD and a huge challenge for its adopters.

On the one hand, the benefits of QFD intended in this way are not limited to the development of successful products but include a structural improvement of company workflows, with many positive side effects related to standardization, like consistency of quality, long-term efficiency (errors are not repeated, the wheel is not reinvented), a more structured decision process ensuring fitness of decisions to company strategy, and employee satisfaction [4].

On the other hand, to reshape the way a company works a broad agreement and commitment are needed, horizontally (over different departments and managers), vertically (from product design team to top management), and in time (the new structure should survive managers turnover and future restructurings). These conditions are not easy to reach and maintain and may limit the number of companies which can benefit from a rigorous top-down QFD approach or slow its adoption.

A broad deployment of QFD through many departments is made even more difficult by the fact that, for most people, the QFD process is not intuitive. Almost everybody can learn notions like the difference between "customer needs" and "product functions", repeat it, and even explain this difference with sound examples. However, very few people will internalize these notions into knowledge, and use them instinctively in their daily duties. On the contrary, since their goal is to design a product, they will instinctively focus on its features (i.e. how it looks or works). A single person in a key position who cannot see the value of QFD and does not understand its principles can make the adoption of the method very challenging.

The top-down creation of a solid QFD process in a company according to ISO 16355 standard is, for instance, part of the mission of the "QFD Institute", which has provided for many years consultancy and training to help companies deploy customer-centric processes. This book is heavily based on the Blitz QFD®, a method promoted by this organization [5], but it contains only a small subset of the tools formalized in ISO 16355 and has a more limited aim and a stronger focus:

- It is dedicated to individuals, not to companies: it is intended to allow employees of industry to use QFD methods for their own daily work, and to promote a bottom-up customer-centric revolution.
- It is dedicated to a very specific phase of the product lifecycle: the product design.

1.3 THE "MINIMUM MERCHANDISABLE PRODUCT"

This book presents several tools which can help to translate the voice of customers into the design of a "minimum merchandisable product.'" In this context, a "minimum merchandisable product" is a product that:

i. Satisfies the top unmet needs of most important customers.
ii. Has a price lower than the product value for the customer (the cost of the pain).

iii. Has a price higher than its cost for the company.

iv. Compares favorably against competition.

The method we are outlining does not consist simply of understanding the customer's needs and translating them into a meaningful product design. This would be completely unrealistic, since in practical circumstances you cannot satisfy all the needs of all customers in view of budget and time constraints.

Instead, in order to beat the competition, it is essential to focus on the most essential product features and resist the temptation of overengineering. These essential features are the solution to the top needs of most important customers, and their development requires a continuous focus with priorities to decide what to do and what not to do. For instance, you need to decide on which customer segment to focus on first, which customers in this segment to talk to, which needs of these customers must be satisfied and which can be left unsatisfied, which is their optimal level of satisfaction, etc. Consequently, you must be ready to neglect many needs of many customers that are not as high a priority, take the responsibility of this recommendation, and be able to explain why this is the best decision.

1.4 CUSTOMER-CENTRIC DESIGN IN PRODUCT LIFECYCLE

The role of customer-centric design in product lifecycle is explained in Figure 1.1. The cornerstone of this process is market research providing a good understanding of customers (via an investigation of their needs) and market (with a competitive analysis of products offered by competitors). This amount of information creates the basis to define the product concept, including both functional (what the product does) and nonfunctional (how functional requirements are implemented, technologies are adopted to develop requirements, etc.) product features.

Market research is mostly "descriptive" since the needs of customers are not under your control and your role is to understand them by introducing the minimum possible bias. "Descriptive" does not mean anyway "passive": you will have to work hard

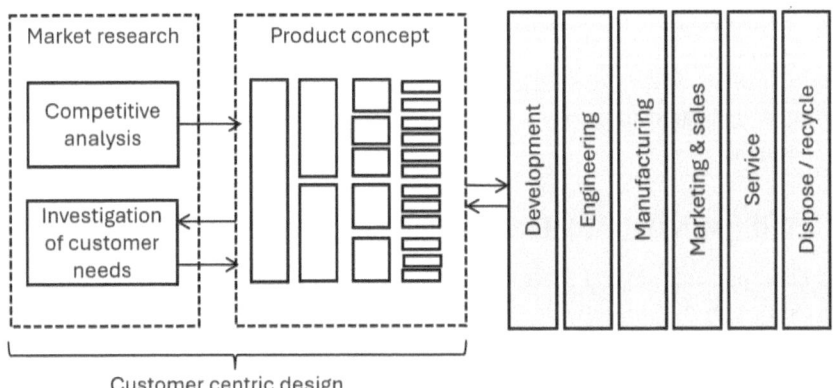

FIGURE 1.1   The role of customer-centric design in product lifecycle.

to truly understand the customers, and possibly help them to develop full awareness of their unconscious needs. If properly conducted, market research provides a solid basis to the "artistic" stage of conceptualization, which is the right place for the creativity of the design team to find satisfaction. Note that design outcomes must always be double-checked with customers in order to be sure that they provide a solution to their needs.

Product design is just the beginning of a product lifecycle. The product will have to be developed in all its details and components, then manufactured (which in case of hardware products requires the procurement of parts), commercialized, and serviced. This volume however focuses on market research and product design for two reasons:

- In initial phases of product lifecycle, customer centricity is critical. A misalignment between product features and customer needs during the design stage can be fatal to the product and almost impossible to correct later.
- Customer-centric design is based on general principles which can be used in many different cases, like software and hardware, products of manufacturing, and services. The subsequent steps in product lifecycles are more product specific and would deserve dedicated discussions [6].

1.5   WHAT YOU WILL FIND IN THIS BOOK

The main steps of the proposed customer-centric product design method, which will be explained in detail in next sections, are shown in Figure 1.2. These steps cover:

- A clear definition of project goals.
- Optionally, a preliminary study based on public information about competitive landscape (relevant products which have been announced or are already in the market) and customers (publications, documentary, etc.). This information is likely to be already available in the company, at least implicitly. Be sure that it is made explicit because it will be a valuable guide for all subsequent steps.
- Prioritization of customer segments.
- Interviews with and visits to customers to understand their situation and pains (the "voice of customers").
- Formalization of the voice of customers into prioritized customer needs.
- Definition of the requirements necessary to design a product which satisfies the most important customer needs.
- Proof of the design against competitive landscape (competitive analysis).
- Proof of the design with customers and internal stakeholders.

As shown in Figure 1.2, there are steps that are carried with customers (e.g. hearing their "voice"), steps that are performed mostly internally (e.g. definition or requirement is mainly the task of the design team), and steps that involve both customers and designers (like the definition of customer needs). The vertical size of boxes in

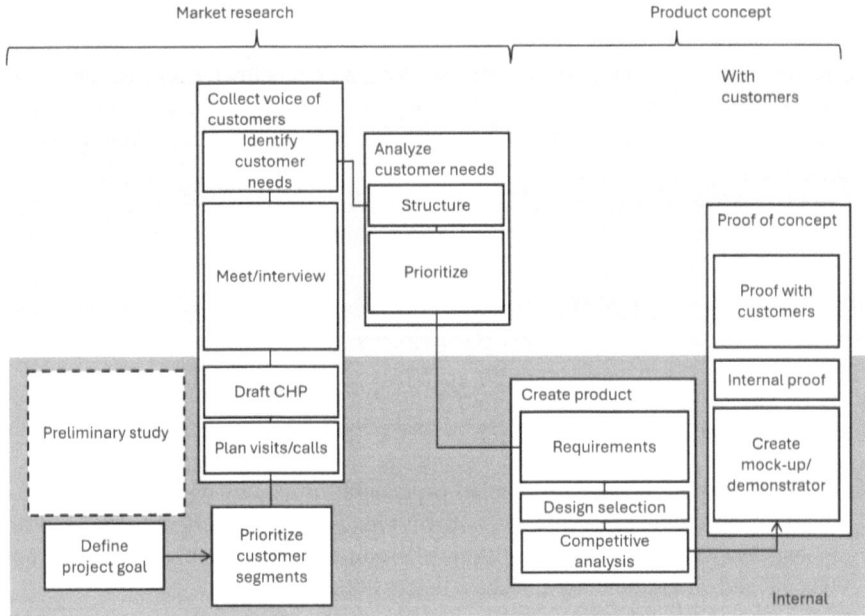

FIGURE 1.2   Main stages of customer-centric product design.

Figure 1.2 is a rough, qualitative indication of the efforts required by every step; the subject of design efforts will be discussed in more detail in Chapter 11.

It is important to note that Figure 1.2 illustrates an idealized, end-to-end process for a new product. This process will have to be adapted to your role and to the nature of the product. For instance, if you work in marketing, you are likely to be more interested in the steps up to prioritization of customer needs and to competitive analysis, while if you are in R&D you are probably more focused on requirements and proof of design. Similarly, to update an existing product with an established market position many steps on the left of Figure 1.2 may be simplified or omitted. Despite all this variability, once you have understood the method and its philosophy, you will be able to jump in and exit at any point. Also you will also be able to determine the input data you need, the tasks you must accomplish, the tools you should use, and the output you must deliver.

Customer-centric design has significant relationship with awareness (what am I doing? for which reason?, etc.) and common sense. The tools presented in the next chapters promote a structured way of thinking which facilitates the development of the awareness necessary to isolate the simplest and more effective solution to a problem. These tools are not rocket science but still require exercise and discipline.

You shall not expect to find some magic formula which takes your problems as an input and generates a solution.[2] Such a solution can be generated only by your intelligence, and the tools presented in this volume will guide and support that activity.

1.6 AN IDEALIZED CASE STUDY

The possibility of adding one or more case studies to this book was carefully considered by the authors. The challenge with case studies taken from real life is that they tend to be complex and industry specific. Many case studies from different ages can be found in the cited literature, but most of them are probably not helping the unexperienced readers to understand the subject better and faster. Therefore, we followed a different strategy:

a. Small excerpts from case studies from the literature, limited to specific steps or tools in customer-centric design, after having been simplified and adapted, are used as examples throughout the book.
b. An artificial and idealized case study has been created to explain the full design process. This idealized case study is based on an episode of "The Simpsons" cartoon series, in which the main character Homer is considered the "average American customer" and hired as a consultant by an automotive enterprise [7] to support the design of a new car. We recommend watching this episode: it is very interesting because the characters have the chance to follow the main rules of customer centricity, but these rules are in the end deliberately broken due to their ineptitude (which is what makes the plot entertaining). This idealized case study will be indicated in the following as "the car for all regular Joes".

1.7 THE STIMULUS–PERCEPTION RELATIONSHIP

In many parts of this book, we will deal with human judgment – for instance, when you will ask customers to prioritize their needs, or when your engineers will evaluate technical solutions suitable to solve customer needs, or when customers will judge value and price of your product when it hits the market.

It is thus very important to understand from the beginning that the relationship between judgment (i.e. perception) and what is judged (i.e. stimulus) is not linear but sublinear.

It is easier to understand this concept by starting with an example. Try this exercise: think about objects of different weight (the stimulus) you may need to handle in your everyday life and judge the efforts you need to lift them (the perception). You may easily generate a scale like the one in Table 1.1. The efforts grades are of course subjective, depending on age, body size, and habits: a backpacker or a construction worker may routinely lift 50 kg without trouble. However, what is important is that to generate a change in perception (i.e. to move from one efforts level to the next one) you need to at least double the weight of the object. Data of Table 1.1 are plotted in Figure 1.3, and two different trend lines are superimposed on data. As you can see, perceived efforts grow less than linearly with increasing weight.

This sublinear behavior is not surprising, since it is known that a logarithmic relationship between perception P and stimulus S is suitable to describe several types of

TABLE 1.1

Example of classification of objects from daily life according to their weight

Examples of objects	Typical object weight	Efforts needed for lifting	
		Level	Description
A pen, A bottle of beer, A smart phone	10–500 g	0	Can be lifted with one hand, without efforts (still a weight is perceived)
A carton of juice, A laptop	1–3 kg	1	Can be lifted with one hand, with some effort
A chair, A 5-kg bag of potatoes	5–7 kg	2	Demanding with one hand, easy with two
10 g of flour A TV set	10–14 kg	3	Two hands and some efforts are needed
A 7-year-old child A car tire with rim	20–30 kg	4	Two hands and substantial efforts
A wardrobe	40–60 kg	5	Cannot be lifted alone

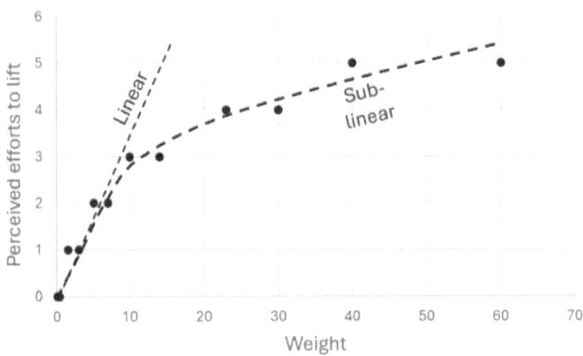

FIGURE 1.3 Plot of "perceived efforts to lift" versus object weight according to the example described in the text.

human perception phenomena like light intensity and weight, a fact that in psychology is formalized by the Weber–Fechner law:

$$P = k \cdot ln\left(\frac{S}{S_o}\right), \tag{1.1}$$

$$S = S_o \cdot e^{P/k}, \tag{1.2}$$

where k is a constant specific to the described phenomenon and S_0 is the minimum stimulus necessary to create a perception. Note that the perception vanishes when the stimulus approaches the threshold S_0.

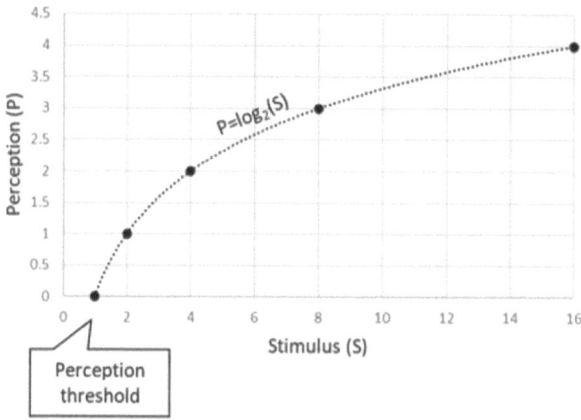

FIGURE 1.4 Example of logarithmic perception–stimulus relationship ($2n$ geometrical progression). Discrete levels obtained by setting $P = 1,2,3,...$ are indicated with markers.

One of the simplest scales which can be directly derived from Weber–Fechner law is the geometrical progression obtained by setting $S_0 = 1$ and $k = 1/\ln(2)$ in Equations (1.1) and (1.2):

$$P = log_2(S),$$

(1.3)

$$S = 2^P.$$

(1.4)

This P(S) relationship is plotted in Figure 1.4 (dotted line).

Often you ask someone to provide their judgment by choosing between a set of discrete levels. This situation can be described by setting $P_n = n$, $n = 1,2,...$ in Equation (1.4). The corresponding stimulus levels are:

$$S_n = 2^n.$$

(1.5)

This scale is plotted in Figure 1.4 too (markers).

Empiric versions of the Weber–Fechner law are used in many fields. For instance, most salespeople believe that you need a strong change in quality or price to stimulate the customer to buy something.

NOTES

1   The same concept is very often indicated as "minimum viable products" or MVP, although the latter term was introduced in the contest of "lean startup" as "that product which has just those features (and no more) that allows you to ship a product that resonates with early adopters; some of whom will pay you money or give you feedback" [8].

2   This would be the effect of a bias known as the "illusion of potential", which will be discussed further in sec. 12.7.

BIBLIOGRAPHY

[1] International Organization for Standardization, "Application of statistical and related methods to new technology and product development process- Part 1: General principles and perspectives of Quality Function Deployment (QFD)," ISO Standard No. 16355-1, Geneva, 2015.

[2] Akao, Y., "An introduction to quality function deployment," in Y. Akao, editor, Quality function deployment, integrating customer requirements into product design, translated by Glenn H. Mazur. Productivity Press, New York, 1990.

[3] Akao, Y., Harada A., Matsumoto K., "Quality function deployment and technology deployment," in Y. Akao, editor, Quality function deployment, integrating customer requirements into product design. Productivity Press, New York, 1990.

[4] Miguel, P. A. C. "The state-of-the-art of the Brazilian QFD applications at the top 500 companies," International Journal of Quality and Reliability Management, vol. 20, no. 1, pp. 74–89, 2003.

[5] Mazur, Glenn H. "Blitz QFD- the lean approach to product development," presented at ASQ World Conference on Quality and Improvement, Anaheim (CA), 2012.

[6] Mazur, Glenn H. "Integrating QFD into phase-gate product design", white paper, QFD Institute, 2010.

[7] Groening M., Brooks J. L., Simon S. (Writers), Archer W. (Director), aired 1991-02.21, "Oh brother, where art thou?", season 2, episode 15, Adamson L. J., Brooks J. L., Groening M., Kogen J. (Producers), "The Simpsons", Gracie Films, 20th Century Fox Television.

[8] Mizuno S., Management for quality improvement: The 7 new QC tools. Taylor & Francis, 1988.

2 Sources of Data

In the customer-centric approach, market data are used to provide the foundation of design, and to prove its quality. Since data collection plays a fundamental role at every stage of this method, it is worth discussing as early as possible various sources of data and the evidence they can provide.

2.1 DATA SOURCES AND LEVEL OF EVIDENCE

High-quality data require more effort to be collected, but the results of their analysis will have strong evidence. On the contrary, data collected with moderate effort will have a lower quality and will support conclusions with weaker evidence. This trade-off is intuitive and can be supported by many examples, like evidence-based medicine [1]. Data sources relevant to our purpose are shown in Figure 2.1, where they are ordered according to their evidence. Sources on top provide the strongest evidence, but they are at the same time the most resource demanding; sources at the bottom of the stack provide lower evidence but are cheaper. Every source of data is characterized by several uncertainties, and it brings a specific value into the design process, as shown in Figure 2.2.

Investigations based on data sources in the upper half of the stack are "designed". This means that you organize data collection, you have control over which customers you meet and the questions you ask. They include, from top to bottom:

- "Gembas", a set of visits to customers during which you can observe them performing the activities your product is intended to support, at their place. In Japanese the term "Gemba" indicates a "crime scene, a place where evidence can be discovered. A Gemba is very different from an interview, since the main goal is not to talk to customers, but to observe them and deploy all your senses (sight, sound, touch, taste, smell, and cognition) to understand the situation. In contrast, discussions are limited to sound; surveys and questionnaires are limited to cognition. Gemba visits require much effort but are nevertheless so valuable because they allow you to collect the maximum amount of information,

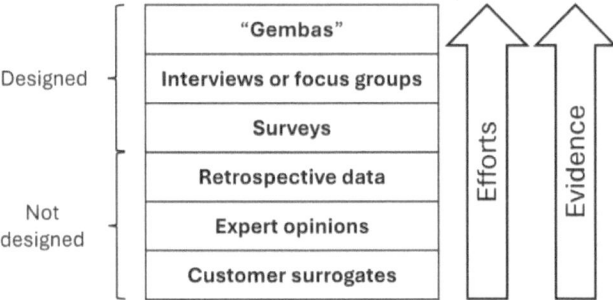

FIGURE 2.1 Sources of data relevant to customer-centric design. Sources on top (typically designed) provide stronger evidence, but their collection requires more effort. Sources at the bottom (typically not designed) are easier to collect but provide weaker evidence.

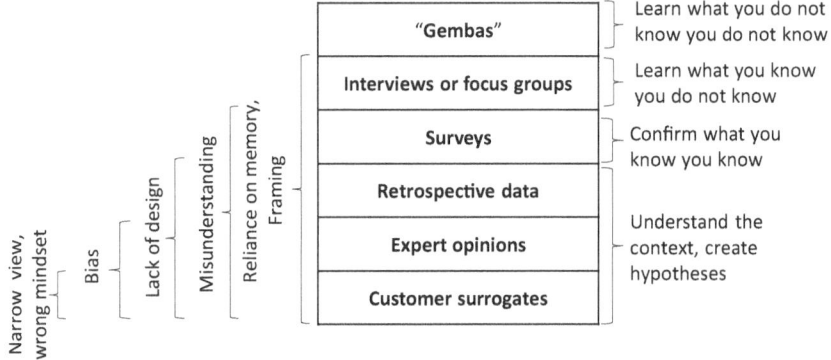

FIGURE 2.2 Uncertainties (left side) and value (right side) of the data sources discussed in this section.

and to receive data beyond what you wanted to ask about. In particular, going to Gemba is the most effective way to discover unspoken customer needs, which means you can learn "what you don't know you don't know".

- Interviews and focus groups (a group of customers brought together out of their place to discuss your subject). These sources are less reliable than Gemba because you will rely on customer memory rather than observing customers performing their life or work activities, and because what you hear will be partially framed by your questions. Here you can learn "what you know you don't know".
- Surveys. They introduce an additional uncertainty: you can never be sure that the customer interpreted your questions in the same way you do. Thus, there is a significant risk of misunderstanding. Surveys are however suitable to confirm "what you know you know" if, for instance, you can use closed-ended questions.

Studies involving the data sources in the bottom half of the evidence hierarchy are "not designed". This means that your control on data sources is limited, and you will use the data collected by someone else who probably had a purpose different from yours. Nevertheless, these data sources are suitable to understand the context and to create hypotheses. Additionally, they can reveal needs that are known to the expert but ignored by most customers, like the need to comply with a new regulation which was recently issued, or which is going to be. In order of decreasing evidence, we have:

- Retrospective data, like books, journals, and documentaries.
- The opinion of an expert, who will rely on their personal experience. Being an expert, they will have a general view and a broad knowledge of their peers. However, the information they provide you during a conversation will have a lot of biases and the level of evidence they offer will be typically lower than, for example, reviews published in high-reputation journals. Of course, reviews are written by experts, but the process of writing and peer reviews help to remove biases, as described in a Chapter 12.
- Customer surrogates between your colleagues. They could be, for instance, former customers who now work in your company, or colleagues who are constantly in contact with customers (like sales force or customer service). If your company sells goods to the public, like cars, clothes, or appliances, every colleague (or even you) may be a customer as well. In all these cases, one must be very careful. For instance, if a colleague is a former customer, and even if the recollection of their experience is intact, they probably left their "customer position" before they could have become experts in the field, and thus may not possess a broad view. Besides that, their experience is now interpreted with a mindset different from that of customers and they are probably biased by your company culture.

2.2 RELEVANCE OF DATA SOURCES

Although different sources of data offer different levels of evidence, they all are important and play a role at some point in the design process. This is possible because their cost decreases with decreasing evidence. It is the same situation in which a low-quality product can find a place in the market if it is of low price. This is explained in Figure 2.3, which shows an exemplary journey along the product design process. Product design typically develops through three phases:

1. During the preliminary study, you use resources available to the public, like published data, possibly integrated with the feedback of experts in contact with the company.
2. Customer interviews, focus groups, and Gemba visits are the sources necessary to collect the voice of customers, translate it into customer needs and prioritize the needs.
3. Interviews, focus groups, and possibly surveys are the most efficient sources to prove that the product design you have conceived is the actual solution to

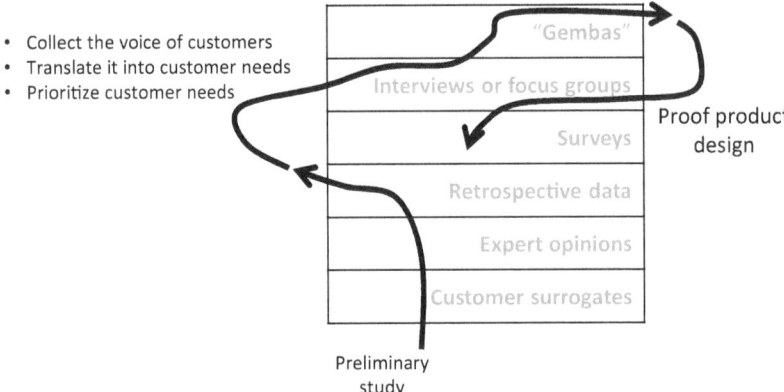

- Collect the voice of customers
- Translate it into customer needs
- Prioritize customer needs

"Gembas"

Interviews or focus groups

Surveys

Retrospective data

Expert opinions

Customer surrogates

Proof product
design

Preliminary
study

FIGURE 2.3   Possible use of different data sources in the product design process.

top customer needs. In this phase, the risk of misunderstanding is very low because customers can see how the product will look like, and surveys can be important to reach many customers at the same time.

BIBLIOGRAPHY

[1]   Burns P., Rohrich R. J., Chung K. C., "The levels of evidence and their role in evidence-based medicine," *Plastic and Reconstructive Surgery.*, vol. 128, p. 305, 2011.

3 Start the Project

This chapter covers the steps from the beginning of the product design project to the point at which you can start looking for the actual names and addresses of the customers for your market research. These steps are aimed to clarify the goal of the project and the customer segment(s) which is the most important to reach this goal.

3.1 SELECT THE PROJECT

Although the shaping of company strategy through the selection of product portfolio is not the subject of this book, it is important to present a short discussion on project selection because you may have the chance to select the project you are going to work on from among a few options and, in this case, it is worth making some effort to take a wise decision.

It is not sufficient to be skilled to be successful; you must find yourself at the right time, at the right place. Although the place and the time are difficult to foresee and plan, you can try to maximize the likelihood to be there by doing constantly something meaningful, that is, an activity suitable to create value. Be sure then that you consider the following aspects:

- The perceived project value for the company should be as high as possible. The reason is that a customer-centric approach is less intuitive than a product-driven approach. If the project is not important, many colleagues will not spend the efforts necessary to understand what it means. Therefore, the project will be a good candidate to be terminated in case at a certain point in time focus is needed on other projects of higher priorities.
- The potential growth it offers. A maintenance project aimed to refresh an existing product will benefit from a customer-centric approach less than the development of a brand-new product.
- The willingness to change. If you start investigating what is needed by the customers, you must be ready to be surprised too. Customers may question what is considered true beyond any doubt in your company and express a need for something different from what is currently done. Remember the company manufacturing measurement instruments which we mentioned in Section 1.1.2:

the revelation that customers enjoyed already enough measurement precision was almost shocking for them. Nevertheless, the evidence brought by the study was strong enough to convince them investing on what customers really needed: product usability. It is therefore preferable to select a project which will be carried out by a team open to change instead of a very conservative one, since everybody may have to leave their comfort zone.

3.2   DEFINE THE GOAL

A project can start in many ways. It can be that the company wants to follow a market trend, to complete or differentiate a product portfolio, or to address a competitive threat. The management may want to achieve innovative goals like figuring out how to remain the preferred supplier in the future of an evolving market, or how to cultivate a new, brilliant product idea. Or maybe the goal is limited to "the future of" some existing product which is becoming obsolete. In any case, the result of your work shall fulfill some basic expectations. A "goal" is a guide to understand, at any time during the project, if you are proceeding in the right direction at the forecasted pace. The direction is right if, at the end of the path, the expectations of the company will be satisfied.

A formal goal clarification is necessary because expectations are based on what managers know, not on what you are aware of. Since you cannot hope to fulfill your assignment by chance, you must:

- Explicitly state the goal. Product design is probably not the real goal of the project; there is always an underlying business goal.
- Clarify the context in which the goal is stated: you and other internal stakeholders must have the same understanding of the situation. If others know more than you, you must learn. If you know more than others, you must explain. This is the right time to perform a preliminary study by reading a bibliography, talking to some expert, meeting sales colleagues, etc. Be aware of and understand that in certain delicate situations managers may not be allowed to explain the full context. For instance, the product may be needed because of an agreement with a competitor which must be kept secret. They know this information could help you, but it would be too risky to disclose it.
- Have your understanding of project goals approved at a sufficiently high level (e.g. "executive" in a small-to-medium enterprise).

To be even more precise, expectations are based on what managers think they know. In some cases they may not have an explicit understanding of the goal, and decisions are taken without a solid logic supporting them. Maybe it is felt that this product is needed, but it is hard to say why. In this situation, the exercise of stating the goal helps to make implicit knowledge explicit, which will be beneficial at all levels.

When stating the goal, the simpler, the better. In some cases a simple statement may be enough: it can be used as a headline in every presentation, to avoid the risk that the goal changes in the mind of colleagues during the project. This is a good real-life example:

The new product shall contribute to:

- *Reduce volatility with portfolio diversification.*
- *Bring EBITDA (currently about 12%) at least at par with industry benchmark (about 18%).*

These two statements tell a lot: the product should be different from existing ones, possibly aimed to completely different customers, and it should have a high profit margin. It is probably all that is needed to start the project if the context is well understood.

In some cases, a single statement is not enough to determine project direction, and additional data will have to be supplemented. Remember that stating obvious facts dilutes the value of this exercise by giving documentation an unprofessional look and wasting time, space in slides, and audience attention during meetings. Generic statements like "generating a profit", "increasing market shares", and so on are, in most cases, obvious. On the other hand, if the product is developed for strategic reasons other than directly generating a profit, then it is worth investigating in detail this exceptional situation.

There are many "guidelines/good practices" to state goals. They can be useful or not, depending on the situation. A well-known example is provided by "S.M.A.R.T" goals [1]. According to this model, goals should be:

- Specific: which kind of product shall be developed?
- Measurable: how is success measured? Common metrics are order intake, number of product units sold, change in market share, gross or net profit, etc. For instance, which is the current level of, for example, revenues per year for similar products, and how should it improve with the new product?
- Assignable: is it possible to say who shall do what to achieve the goal?
- Realistic: ambitious targets are to some extent positive and contribute to motivating the team to do its best. However, shooting for the moon (which may be inspiring at a company strategy level) is more likely to give the perception inside and outside the team that this project is not serious.
- Time-related: this usually means that there is a deadline for the completion or a set of milestones.

You must be very careful with quantification with numbers like success metric, deadlines, etc. Quantification has a great potential if numbers are reliable. If they are not, or if their uncertainty is too high, numbers just dilute the value of the goal statement, and possibly create a false sense of having under control a situation which is not. Consider the example presented in Table 3.1. Before using this table, one should ask oneself the following questions: is it reasonable to double revenues in 4 years? If it does not happen, will the project be deemed a failure? What is the opinion of the head of product management, the one who will determine whether the project has been a success or a failure?

TABLE 3.1
Example of project goal definition with too uncertain numerical values

Goal	How measured	Current level	Target level	By when?	Who judges success
Increase worldwide sales volume for product line XX with a product aimed at customers who are not potential buyers of our high-end solution	Revenues/y	7.8M€	15.2M€	4y from now	Head of product management

3.3   CUSTOMER SEGMENTATION

Segmenting the market means dividing the potential customers into groups based on some shared characteristics. In this context, we want to understand and prioritize "customer needs". Thus, we shall ideally create homogeneous groups of customers with the same top needs. However, these needs are unknown at the beginning, and we must start with some assumptions that, in the end, may turn out to be incomplete or even wrong. This is not bad, because if eventually you reach a good understanding of a market which is difficult to understand, no matter how good or bad your initial assumptions were, you will be able to design an innovative product beyond the imagination of competition. Remember that the main goal of this initial market segmentation is to decide which are the first customers you want to talk to. If necessary, you can include other customers later.

In many cases you can be pragmatic and start with the typical marketing-based segmentations of your company, which probably reflect not only the market but also the logistic structure of the firm. If you can use the current structure of sales and customer service, at least in part, it will be easier for your colleagues to support you in identifying and approaching customers.

If, for example, your company sells products to other companies, and customers are customarily grouped according to geographical region, size, and products they sell to their end customers, then the segmentation could look like the one presented in Table 3.2.

If you decide to follow this approach and use an existing customer segmentation, review the available segments based on your current understanding of the market for the new product. Make the necessary adjustments until you can clearly explain basic facts like the following:

- Why customers in different segments are expected to have different needs.
- Which is the relationship between your segments and the whole market. Are the segments completely covering the market? If not, why a portion of the market has been put to the side?

As an example, a customer segmentation used by Boon Rawd Brewery, owner of the Singha beer brand, for market research in the United Kingdom is shown in Table 3.2. This study was actually aimed to design marketing messages, not the

TABLE 3.2
Example of customer segmentation based on existing company segmentation for a business-to-business product

Segment no.	Region	Customer size	Customer product type
1	US	Large	Type 1
2	US	Medium	Type 2
3	Europe	Large	Type 2
4	Asia	Large	Type1

TABLE 3.3
Example of customer segments used for market research by Boon Rawd Brewer (Adapted and simplified from Ref. [4])

Who uses this product?	Where?	When? In which circumstances?	Why? To which purpose?
Business people	Restaurant	Lunch	Go with the meal
			Take the edge off a bad day
		Night	Dinner beverage
			Relax
	Pub	Afternoon, after work	Quench thirst
			Relax
			Socialize with friends
			Wait for friends to arrive
Shoppers at market	At home	Any time	Go with the meal
			Relax
	Parties with friends	Barbecue	Go with the meal
			Socialize with friends

product itself. Nevertheless, its customer segmentation is clean and intuitive and constitutes a very good example. The QFD team identified with local agents two main customer segments: business people, usually drinking beer at pubs and restaurants, and shoppers at market buying beer for domestic consumption or parties with friends. For business people the column "When" indicates the right time for interviews. The column contains initial hypotheses about the motivation to buy the product, to be confirmed by conversations with customers.

If the available customer segmentation is not suitable to the current project, you need to prepare one from scratch. This may be the case if you are exploring new products for existing customers, new markets for current products, or both new markets and new products. A possible procedure [2] consists of listing the most important customer characteristics as headers in a table (which is then called "Customer Segments table" in QFD language), and their possible values as table cell contents. Headers recommended in Blitz QFD®, which with some adaptations are suitable to cover many practical circumstances,[1] are those shown in the example of Table 3.3. A customer

TABLE 3.4

Example of customer segmentation performed from scratch

Who uses this type of product?	Why? To which purpose?	What product is currently used?	When? In which circumstances?	Where?	How is the product used?
Customer type 1	Purpose i	Product A	Event 1	Place a	Process 1
Customer type 2	Purpose ii	Product B	Event 2	Place b	Process 2
Customer type 3	Purpose iii	(none)	Event 3	Place c	
Customer type 4	Purpose iv				

segment can be obtained by choosing one element from every column. An example is given in Table 3.4. The selected top customer segment is composed of the "Customer type 2", using the product for "purpose 3", which is currently achieved with the existing "Product B" which is used at "event 2" in "place a" following "Process 2". Following the same procedure, other customer segments of lower priority can be constructed using the same table by placing additional "circles" on it.

In the simplest case, one segment will clearly be much more important than all the others, and it deserves your focus. In more difficult cases, you will have to consider two or three segments at the same time. Regardless of the method you use to determine customer segments, if you consider more than one, then they must be prioritized. Prioritization is essential and cannot be skipped because working on several segments at the same time will dilute your efforts and it can happen that you are not able to deliver the expected quality in time unless you focus on one or the few with highest priority.

You will discover very soon how reliable your market segmentation is: when you start collecting the voices of customers you will see if they are providing homogeneous feedback. At that point it will be clear if you can go ahead with the existing segmentation, if two segments can be combined, or if one must be split into several. In the latter case, it may be necessary to introduce new customers in the study, to have good statistics in all segments, or to drop some segments to preserve your focus.

An example representative of our idealized case study is shown in Table 3.5. The column "where" has been omitted, since the project is focused on US domestic market. The column "who" is duplicated to consider more demographic parameters relevant for car preferences (gender, family status, and social hierarchy). In the end, the top customer segment is composed of middle-class fathers, who typically use a sedan daily to go to work and have an angry driving style.

You can make customer segmentation quantitative by assigning a weight to customer segments based on their ability to achieve the projects goals. These weights

TABLE 3.5
Example of customer segmentation for the "regular Joe's" project

Who uses the car? (gender & family status)	Who uses the car? (social hierarchy)	Why? To which purpose?	What car is currently used?	When?	How is the car used? (driving style)
Male, single	Working class	Go to work	Sedan	Daily	Angry
Male, with family	Middle class	Go shopping	Station wagon	Weekly	Reckless
Female, single	Upper class	Going to vacation	SUV	During holidays	Anxious
Female, with family			Sport car		Careful

could be based on segment size, customer motivation to buy or willingness to pay a higher price, influence on other segments (one segment may, for instance, include key opinion leaders), etc. This is possible using the prioritization methods against multiple decision criteria, explained in Section 10.2 (see, e.g., Ref. [3]). A quantitative approach can help you if you have a good understanding of the market, but it is too complex to perform a prioritization "at a glance". Quantitative prioritization will take additional efforts to perform the evaluation and to explain it to your colleagues; thus, do it only if you think it brings a real benefit, or if it is explicitly requested. Consider that:

- The only benefit of quantitative segmentation could be your improved awareness.
- You can reach out to only a limited number of customers, and the finer details of your results could be lost.
- Ad hoc corrections will usually be added to include some customers which, despite calculation, appears very important. This could frustrate the efforts spent in quantification.
- Customer acceptance of your contact requests could jeopardize your calculation. In the end you talk to the customers who are willing to talk to you, and customers in one segment may be much more available than in another one.

NOTE

1   The idea of effectively exposing facts by explaining *quis* (who), *quid* (what), *cur* (why), *quem ad modum* (how), *ubi* (where), *quando* (when), and *quibus adminiculis* (by what means) is known and used since the classical age and is commonly called "the seven circumstances" [5].

BIBLIOGRAPHY

[1] Doran, G. T. "There's a S.M.A.R.T. way to write management's goals and objectives," *Management Review*, vol. 70, no. 11, pp. 35–36, 1981.

[2] Childs, D. et al., "DREAM/QFD to re-design staff service excellence at Rutland Regional Medical Center", *Transactions from international symposium on QFD*, p. 51, September 24–25, Portland (OR), 2010.

[3] Ronney, Eric, Peter Olfe, and Glenn H Mazur. "Gemba research in the Japanese cellular phone market," *Transactions from international symposium on QFD*, p1(17), Michigan, USA, 2000.

[4] Vongpatanasin, T., Mazur G., "Thai brewery deploys QFD tools to tap into consumer motivation," ASQ, The Global Voice of Quality, December 18, 2012.

[5] Sloan, M. C. "Aristotle's Nicomachean ethics as the original locus for the Septem circumstantiae." *Classical Philology*, vol. 105, pp. 236–251, 2010.

4 Collect the Voice of Customers

With customer segmentation you have decided which customers are the most important for the project. Now you must approach these customers and understand how a new product could improve their life.

4.1 WHO SHOULD YOU TALK TO?

If your product is sold to the end user, the customer is a person who will decide how to invest their own money to receive a direct benefit. In this situation it is clear who you should talk to. In other cases, however, more people could be involved in the acquisition process, for instance, if the product is not intended for retail customers, but for companies or institutions. In these situations, it may not be trivial to identify your spokesperson at the customer site.

In general, before calling the customer to arrange a visit or an interview, you must have a clear understanding of which voices must be heard. If you are unsure about this point, it could be worth to sketch the so-called "customer value chain" [1]. To do this you must:

1. Identify stakeholders throughout the product lifecycle, including storage, packaging, and disposal.
2. Determine how stakeholders are related to each other.
3. Identify value propositions.
4. Analyze the chain to identify critical spokespersons.

Consider as an example a piece of equipment for radiation therapy. In radiotherapy clinics sophisticated tools are used to treat cancer with great accuracy; radiation oncologists decide how to treat patients and supervise the treatment process, but equipment is maintained and operated by radiotherapy technicians, under the supervision of medical physicists. Thus, the equipment will probably be operated by a technician, but the purchase will be recommended from a physicist to a medical doctor, who will request the buying to the clinic management. All these people will have to feel confident with the product. Additionally, if the product is going to be used in direct contact with the patient, clinic staff must be sure that patients will accept

DOI: 10.1201/9781003544845-4

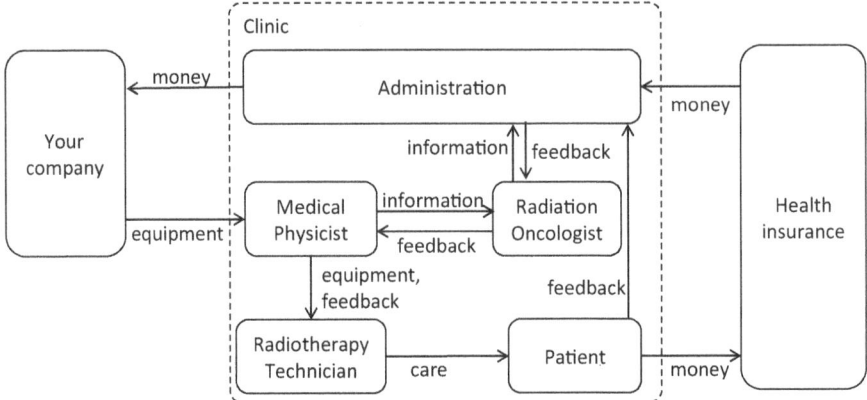

FIGURE 4.1   Customer value chain example: acquisition of equipment for a radiotherapy clinic.

the product too. The customer value chain in this example is shown in Figure 4.1. Stakeholders are represented by boxes, and value propositions are represented by arrows connecting the boxes. Since every stakeholder is somehow paid to provide a benefit, every box must have at least one input and at least one output connection, and you must be sure that the product will not block the flow of value. It takes only one adverse stakeholder to torpedo a sale!

Such a diagram does not need to be an accurate description of reality, otherwise it would become overwhelmingly complex and of limited use. It should capture the most critical connections instead. In this example, one should probably talk to medical physicists in all the clinics included in the market investigation. In a few clinics it will be necessary to meet technicians and physicians too, to be sure that all perceive a value so that the former will not generate resistance to the product practical adoption, and the latter will be willing to support the acquisition in front of management.

Remember that everybody is very sensitive to feedback from people who can judge them. In the radiotherapy example, clinic administration is very attentive to patient feedback which could be represented by positive trends in social media or, in the worst case, by a lawsuit. Based on this data, they will provide feedback (a judgment) to physicians regarding the quality of their work.

4.2   HOW MANY CUSTOMERS TO START WITH?

The more customers you interact with, the better your understanding of the market, but the larger the amount of time and effort you will need to invest. Therefore, the number of customers involved in your study should ideally be the minimum needed to get a good enough understanding of customer needs.

In the initial part of the market research, you deal with customers to understand their life, problems, and needs. This requires mostly the collection of non-numerical, unstructured data, like the customer's free speech and answers to open questions, or your impressions and observations. This kind of research is well known in sociology and healthcare as "qualitative research".

Later, you will deal with prioritization of customer needs and with the validation of your design. Customers will be asked to answer closed-ended questions and possibly to quantify their feelings. The investigation will become, at least to a good extent, "quantitative research".

The number of customers necessary in a qualitative research is typically smaller than that needed in a quantitative analysis. Thus, it does make sense to start with the number of customers necessary for the qualitative part and enroll additional ones later, whenever necessary. This section deals with the efforts necessary for the qualitative part (understanding of customer needs). A possible increase of sample size for the quantitative part will be discussed in later sections.

Well-prepared qualitative research has a learning curve with a favorable shape, that is, a steep initial slope (Figure 4.2, left). This means that you learn a lot in the beginning with moderate effort. The learning curve is very steep close to the origin if:

- You already investigated the context. Thus, you know in which conditions customers operate and you can understand their language. This will minimize any painful initial barriers to overcome. It is like approaching physics if you already know mathematics. Of course, understanding the context costs time, but preliminary research is not the time of your customers, which is your most valuable asset.
- You have developed the mental attitude necessary to proficiently listen to the voice of the customers. This subject is discussed in detail in Chapter 12 dedicated to cognitive errors. Being ready to listen implies humbleness. You should recognize that, no matter how good your preliminary investigation was and how deep your experience is, you can truly understand customers' problems only if you interact with them. It is like bringing an empty bottle with you, which customers will quickly fill with knowledge.

If you are not well prepared, you will have to overcome an initial barrier (Figure 4.2, middle) and some of the time spent with customers will be not as productive. This is a

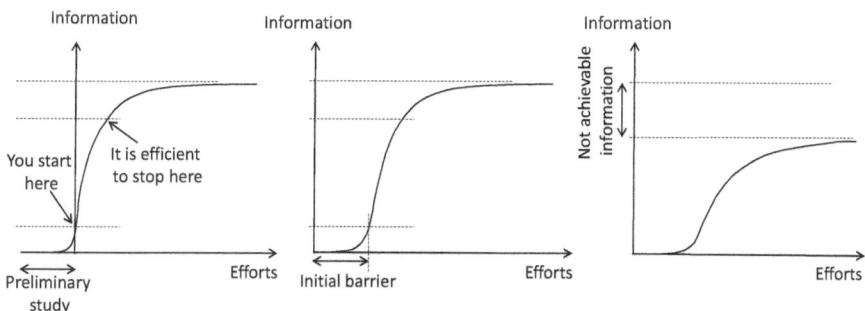

FIGURE 4.2 Possible learning curves during market research.

pity, since the time customers can dedicate to you is limited and should not be wasted. Moreover, customers are typically not the best resource to teach you the basics.[1]

Even worse, if you are not mentally prepared to hear the voice of customers, you will face the situation of Figure 4.2 (right), and you will not be able to fully access the information customers are disclosing to you. The portion of information which remains hidden, the less trivial to learn, is the most likely to contain unspoken needs which could pave the way to innovation.

A common feature of these learning curves is that, at some point, they start to flatten, tending toward a saturation value[2]: initial efforts generate much more information than the same efforts at a later stage. In practice, this means that, at some point, new customers will spend much of the interview time telling you what you already heard from other customers, simply because you already know a lot. From this point on, it is more effective to stop collecting data and to start with data analysis.

The ideal number of qualitative visits is not fixed in general. However, experience suggests that after 10–15 customers you typically see the customer's process, behaviors, and words repeat, indicating that you have probably exhausted that customer segment. This agrees with existing literature about qualitative research. A recent review study [2] revealed that typically 9–17 interviews or 4–8 focus group discussions are needed to reach saturation of knowledge, if the investigated population is homogenous, and the objective of the study is narrowly defined. This is a fraction of what is typically recommended by independent market research agencies which are accustomed to the statistical validity rigor of quantitative studies, and which arguably play on the safe side which ensures quality of results and higher revenues.

4.3   SET UP APPOINTMENTS WITH CUSTOMERS

At this point you have an estimate of customer segments, and how many customers you should contact in the top segment (or in the few top segments). Your next step is to identify the customers who fit these characteristics and are willing to talk to you, and to set up appointments with them.

Often, the easiest way to contact customers is through your sales department; these colleagues know the customers and can recommend those who are more likely to accept a visit or an interview. The sales team may find that your market research is useful for other business reasons too, like strengthening a relationship or contacting again a customer who has been off their radar for a while. The drawback of this approach is that it is focused on the existing customers of your company.

In most cases it is desirable to include customers of the competition too because they may become your customers if you solve their needs. They could be reached, for instance, due to their personal acquaintance with you or other colleagues. When contacting these customers just be sure that they are not too close to the competition, otherwise your plans may leak out to them. Whoever you contact, it is good practice to keep the sales team informed.

Be prepared to be patient, since not all contacted customers will answer, and, even if they do, their answer can be negative. To give an example, according to the personal experience of one of the authors, in the medical field you can expect that about 40% of contacted customers will accept hosting a visit, and a significant part of them will do it only after receiving a reminder. Most of this 40% will agree to review and score needs later, too, but only a fraction of the 40% will accept a pre-visit call (which may be helpful to align on the customer process model, as explained later).

Remember that the time of customers is your most important asset. Customers are not under your control; thus, you must wisely use the time they dedicate to you, trying to learn from them as much as possible. This means that (1) you must be at the appointment at the right time, and fully focused on listening, and (2) after the visit, during data analysis, you must maximize the information which can be extracted from available data. For this reason, it is essential to plan your appointments and minimize the risk that something goes wrong. A plan B is something nice to have too. Few details you should check to properly plan the visit are:

- In general: is all the necessary material at hand? Do you know who the customer is? Do you know basic facts about the customer? Do you expect to meet a technical person or manager? Can you pronounce their name?
- For a visit: are you in the correct time zone? Do you have a phone number? Do you have a time buffer for travel delays? Is someone picking you up? Is a sales-person joining you?
- For video calls: is your internet connection stable? Are files available and at hand on your personal computer? Can you find a quiet place? Has IT department solved any previous problems you had during previous calls?

A very effective way to avoid forgetting something important is to create a table with the essential data about your customer selection. In the QFD language it is called the "customer contact checklist". A spreadsheet is likely to be the best option: then you can dedicate a row to each customer and use as many columns as needed for data. An example of essential columns is proposed in Table 4.1:

1. Customer name and contact(s). Unless you sell directly to private people, for each customer (company or institution) you may have more than one contact. For instance, an assistant may be included in the communication loop. For video calls the contact email account is sufficient, but for visits you need the address and at least a phone number, possibly both mobile and landline, to be sure you can quickly reach out to someone if you have difficulties (e.g., if you are late or if you cannot find the place).
2. Region: the geographical information you need to quickly localize the customer. It could include country, state, etc. It can be included in the contact, but it is useful to have a separate column to sort the table and group together all the customers from the same region.
3. Status: what you already did with the customer. Ideally use many lines, one per event, starting with the date, so that you can easily reconstruct what happened with every customer.
4. Due date for the next step: by when the next step must be accomplished? This column is very important, since you can use it to sort the full table and quickly identify the most urgent actions.
5. What must be done at the next step.

TABLE 4.1
Exemplary table with essential fields that could be used to plan a customer visit

Customer	Region	Status	Next Step	
			Due Date	**What**
Company 1 main contact: name, last name, email, street address office assistant: name, last name, email landline, mobile	UK	**2022.08.04** visit planned **2022.07.15** positive feedback **2022.07.04** visit proposed	2022.09.15	on site visit
Company 2 main contact: name, last name, email	US	**2022.08.22** visit done **2022.08.15** visit planned with assistance **2022.08.03** positive feedback **2022.07.26** visit proposed	2022.11.25	set up appointment for prioritization
Company 3 main contact: name, last name, email deputy: name, last name, email landline, mobile street address	Japan	**2022.08.23** video call cancelled due to deputy unavailability **2022.08.18** video call planned with deputy **2022.08.08** deputy takes over **2022.07.28** visit proposed	2022.09.06	contact deputy again to plan a new video call

Depending on the situation, you may find that additional fields are beneficial. Examples are the following: who is going to meet the customer; the official contact in your company (for instance, the local area sales manager); details about the customer (why are they interesting?); lessons learned; special relationship between the customer and your company etc.

4.4   DATA RECORDING

When meeting customers you need to record in real time your experience, and efficiently store this data and retrieve them later for analysis. You cannot simply rely on your memory, as explained in Chapter 12, because you would be very likely to forget and distort data due to cognitive errors. There are many means to record data, each with its pros and cons. A list of typical recording tools is shown in Table 4.2.

The richest amount of data can be taken with a video camera, in which case you can have an unfiltered record of what the customer says and what happens. Video

TABLE 4.2

Comparison of different tools to record the voice of customers

Recording Mean	Advantages	Disadvantages
Video camera	• Rich data	Prior authorization needed Data protection concerns Watching footage doubles visit time An extra person full time busy with the camera Customers may behave differently if recorded
Audio recording	• Direct recording of VoC • Interviewee authorization is sufficient	Another tool to annotate visual observations is needed People may talk away from the microphone Listening to recording doubles visit time Customers may behave differently if recorded
Audio notes	• Manageable, light • Concise	Voice of customers is interpreted from the beginning Partial keyboard retyping needed
Typed notes	• Data ready in electronic form for further analysis • Concise	Voice of customers interpreted from the beginning Fast keyboard typing needed Waiting for computer to start may be annoying
Handwritten notes	• Manageable, light, • Ease of drawing • Concise	Voice of customers interpreted from the beginning Fast handwriting needed Often you stand up Partial keyboard retyping needed

recording has the highest number of drawbacks too. First, a preliminary authorization to record a video will be needed, which may not be easy to obtain due to concerns with protection of personal data and industry secrets. This concern may be overcome by carefully reviewing the video at the customer's site, using a portable recording media (e.g. an USB stick) and leaving it behind with the customer. Second, video recording maximizes the effort required on your side: a person will be full time busy with the camera during the *gemba*, and you will have to watch the footage at least once to write down the voice of customers: this basically doubles the time dedicated to the visits. A third concern could be that customers may behave differently, and less spontaneously, in front of a camera. Experience has however shown that customers soon tend to ignore the camera as they realize your role is to help them improve their work and not a threat, because you have no authority over them.

At the bottom of Table 4.2, we find handwritten notes. This is an option only for people who can write fast, an ability which is unfortunately often lost after completion of studies. If you fall in this category, this option can be an easy one even if you are meeting the customers alone (which anyhow is not a good practice). An advantage of handwritten notes is that you can easily add drawings. Main drawbacks are that you are interpreting the original voice of customers, which is often wandering and nonlinear, into "summarizing bits" and that eventually you will probably need additional time to retype your annotations on a keyboard.

The best means, or the best combination of means, depends on who is meeting customers and on customer settings. In many cases electronic notes work well with video calls, which could be complemented with a video or audio recording if the customer agrees. For Gemba visits a combination of text notes (handwritten or electronics) and video (for visual data like a demo of how the customer works) may be a good compromise too.

4.5   CUSTOMER PROCESS MODEL

It is important to have a standardized procedure to follow during interviews and visits because of several reasons:

- You have a limited time with customers thus you must know precisely what to do.
- You need to compare and aggregate data from different customers.
- Later you will need to share data with your team.

The minimum level of standardization you should achieve is a predetermined list of questions. This can be sufficient during the first contacts and pilot interviews if questions are well designed and focused on customer's work, and not on the product. However, you should adopt the so-called "Customer Process Model (CPM)" as soon as you have enough knowledge to draft it because it offers a better basis to focus the conversation on the customers and discover their needs.

The CPM represents your understanding of the customer's workflow in which the product will be used. It should be as simple as a linear sequence of few (5–15) consecutive steps, in a single sheet. In some cases the CPM will contain steps which are executed strictly in sequence as part of a work procedure. In other cases, "steps" could represent actions taken at different places or times or, at an even higher level of abstraction, actions which are repeated in a periodic or unregular way. You can also include relevant actions performed out of the workflow (before, after, or in parallel), since customer needs discovered during these stages may lead to additional functionalities increasing the value of the product. You can determine the content of the diagram with great freedom, the only rule being to keep it simple. When you prepare the layout of this diagram, leave enough space around the blocks to take notes since it is very convenient to annotate customer feedback directly on the diagram sheet.

The CPM example of Figure 4.3 represents the process during which a physician relocates to work in a hospital in a rural community. It is taken from a study aimed at

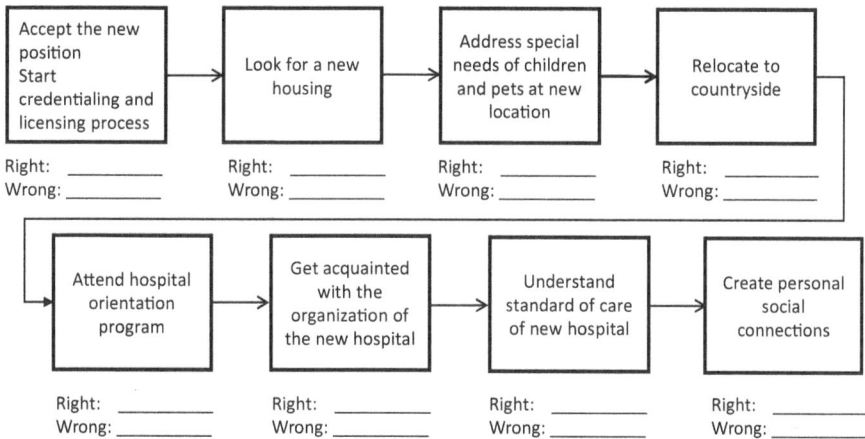

FIGURE 4.3   Example of customer process model (adapted from Ref. [8]).

redesigning the orientation process with which the hospital helps the physicians and their families integrate into the local community. In this example, fields to annotate pains (what goes wrong in the customer's process step), and what satisfies the customer (what is right), have been pre-typed below every step.

The CPM must be reviewed with the customer (before the visit if possible, or at the very beginning of an interview), adjusted to reflect his specific situation and discussed. You must use a different form for each customer. Key questions you should ask to the customer during the review of CPM are the following:

- Does this process reflect your activity? If not, what must be changed? Shall we delete or add steps or change their sequence?
- Which steps generate your major pains? What are they? Pain points identified with the CPM are typically worth a closer investigation during the prosecution of the interview or require an observational study, since they are likely to reveal unspoken customer needs.
- Which step is responsible for your "biggest pain". If you are visiting the customer, you may want to observe the customer while performing this step.
- Which steps give you the highest satisfaction? Why? This inquiry is important because people are forgetful about the benefits they are already receiving, and you must ensure they will not be removed in the new product. Thus, your product will probably have to address important needs already solved by existing products, too.
- How do you measure or monetize your satisfaction or dissatisfaction? What threshold must be met (or not exceeded) to trigger a desire to purchase a new product?

An example of customer process model with annotations taken during the interview of a "regular Joe" in our imaginary case study is shown in Figure 4.4. According

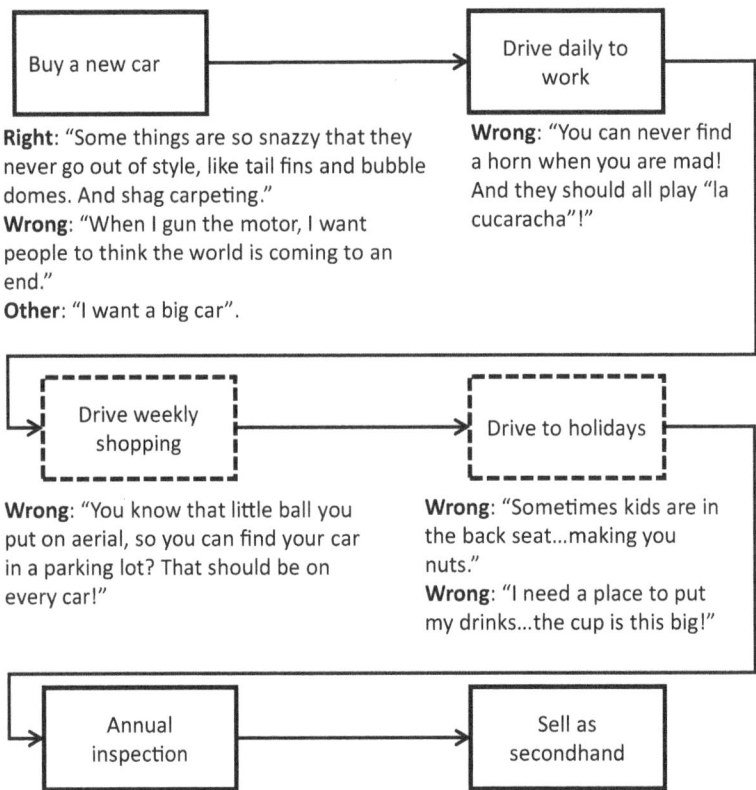

FIGURE 4.4 Customer process model (CPM) with annotations taken during the interview of a "regular Joe like" customer in our imaginary case study. The boxes with solid outline represent the original CPM created by the project team. The boxes with dashed outline have been added following a request of the customer.

to the top customer segment definition for this study, the team created the CPM with a focus on "drive daily to work". However, this specific customer asked to add the steps "drive weekly shopping" and "drive to holidays", which are for him a major source of pain.

You can ask customers to help you generate the CPM, but you should not let them do it alone: they could easily create overcomplicated diagrams trying to provide an accurate description of their situation. An example of the structure of the CPM prepared by a highly educated customer who wanted to accurately describe the complexity of their job, is shown in Figure 4.5 (text has been removed from cards due to confidentiality reasons). This diagram could well serve other purposes, but it is not suitable to guide a conversation aimed at discovering customer needs. Remember that the goal of the CPM is not to precisely describe the workflow in which the product will be used, but to capture its most important aspects. It must be accurate enough, but not overly accurate.

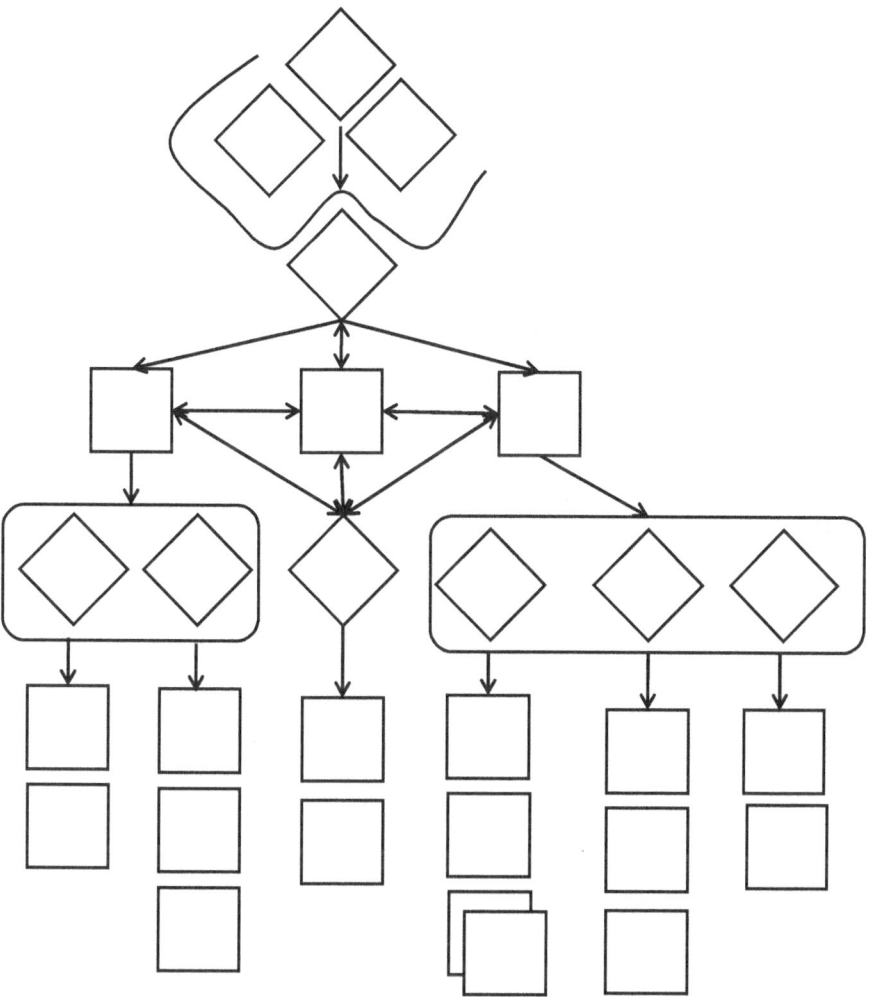

FIGURE 4.5   Structure of overcomplicated customer process model prepared by a customer with insufficient guidance.

Be careful to not drive the conversation away from customer's activity. Specific questions of the team must be reserved until the end of your visit. Questions which could be meaningful at the end of the interview are those about customers' perceptions of existing products from your company or from competition. Be careful however not to turn the Gemba visit into a sales call or a technical service visit. Salespeople can be especially interested in pushing existing products since often their pay is contingent on a sale. Remind them that the purpose of the Gemba visit is to create *tomorrow*'s sales by designing a new product customers want to buy because it improves their work, productivity, and life. Thus, do not promote other products unless the customer explicitly asks for this kind of information. Similarly, technical people may want to validate product concepts they have already in mind with customers. In this case, make them aware that customers are typically not well enough versed in your future technology to give useful input, and even if they agree that your ideas have merit, that does not mean they will actually buy it in the future.

You can of course go further with standardization and define the steps to be performed during visits and interviews with the desired level of detail. A finer standardization may be worth the effort if you are investing many resources in this market research, for instance, if you are going to visit many customers worldwide with a small team. A good example developed by Sandvik Coromant, in which the procedure to be followed during Gemba visits was described with cartoons, is shown in Ref. [3].

4.6   CUSTOMER VISIT/INTERVIEW TABLE

Since in the end you want to compare the feedback of different customers belonging to the same segment, it may be helpful to arrange the data you collected during several visits in a table in which every customer takes a row. You can put on the table the most interesting answers to standardized questions, or most important annotations from the CPM. The number and content of columns depend on your situation; a typical selection includes, beside the customer's name or description, the following:

- Verbatims, for example, memorable statements you heard from customers, which you want to preserve textually.
- Your observations, which may include objective visual observations as well as what you think you have inferred from the situation. This field is particularly important if you perform Gemba visit, during which all six senses may be employed to capture an in-depth understanding of the customer's pain points.
- Clarified items (subjects you have discussed with the customers to the point of reaching a common understanding). Clarified items should state a single issue (do not use "and", "or," or "but"-type conjunctions) and may include how the customer measures them and what their desired performance or outcome would be. Clarified items will be very useful later to extract customer needs from your observations.

An example of one row in the customer visit table, freely adapted from the study of beer consumption [4], is shown in Table 4.3. This row summarizes the observations of the market research time when interviewing a group of nine people having a buffet during lunch break. Note the interesting comment about Thai style food, which is not directly related to the product (the beer) but which reveals an emotional bonding with the country where the beer is brewed. A table which could have been generated in our imaginary regular Joe's case study is shown in Table 4.4.

TABLE 4.3
Example of entry in a real-life customer visit table

Customer	Verbatims	Observations	Clarified Items
A group having buffet: eight men and one woman	"Usually, I have a beer at lunch; I prefer Corona because it is less gassy." "Cook Thai style at home for a special meal that is unusual and romantic (and it works!)."	They need to be in appropriate mood for work (not drunk) after lunch.	They want to make a good use of the one-hour break they have to eat. This time includes covering the 10-min walk distance to the restaurant.

Source: Adapted from Ref. [4].

TABLE 4.4
Example of customer visit table for the regular Joe's case study (inspired by TV show Ref. [7])

Customer	Verbatims	Observations	Clarified Items
Regular Joes	"I want a big car."	Unhappy with models in car dealership	Currently he needs to use his wife's car for shopping.
	"I need a cup holder for my 2-liter water bottle!"	Shows cup size with both hands	Drinks in big cups has a lower cost per unit volume.
	"I need a searchlight attached to my car so I can find it in a parking lot at night. Every car should have that! "	Very emotional, angry, and shouting	Fear of not finding the car when he is out with the family is for him a source of stress.
	"What I like never goes out of style! We need more chrome, and turf carpeting."	He thinks he is very refined	A car contributes a lot to the image of a man.
	"I need a horn button I can smash when I'm mad! And it should sound like [a] gun fire"!"	Explained while dancing	When livid, he needs to express his feelings to calm down. A polished look helps to impress others.
	"When I drive by, I want people to think Armageddon." "My kids are in the backseat…driving me nuts."	Mimic driving, reproducing engine noise with his voice	Engine rumble is the most effective way to intimidate other drivers.

4.7 HINTS AND WARNINGS

Your success in collecting a meaningful voice of customer, from which you can extract customer needs, will improve with time and experience. You will need to develop or refine the soft skills necessary to interact with customers and, at least in the beginning, you may find the following hints, which the authors have derived from their experience, useful:

1. Humans do not like their mistakes highlighted: thus, do not criticize or instruct customers, just watch and listen. One caveat is when the customer may engage in unsafe behavior, and you may need to prevent that.
2. Let the customer do the talking. If possible, try to develop empathy for him, to better understand their feelings and to see reality from their point of view.[3]
3. Be perceived by the customer as receptive and ready to learn. Playing a subordinate role will help, without going so far as to appear incompetent. The customer should feel that you can understand them, and that you are willing to do so. Pay attention to your nonverbal cues (yawning, for instance, will not help developing an effective conversation).
4. At every moment, it should be clear to the customer that they are talking about their experience to one person. Thus, limit one person to be the interviewer and do not create confusion with debates among your team members.
5. Customer's behavior changes when "the boss" is present. The obvious reason is that everybody is very careful of their own behavior in the presence of the person who will judge their professional performances. Ideally you want to hear the unbiased voice of all the people who play an important role in the use and purchase of your product. Unless the user and the one who will approve the purchase are the same person, it is good to talk to both. Ideally you should talk to them separately. If they show up together during a visit, a good strategy could be that a member of your team entertains "the boss" with a more high-level discussion, while someone else interviews and observes the user. If the boss and the user are present during the same video call, you must rely on the art of isolating their points of view with focused and smart questions.
6. Look around and try to identify customers' troubles. Problematic situations which are not explicitly perceived as troubles (for instance, because the customer is used to them, or they relate to a product you do not provide) are an interesting gateway to unspoken needs. As an example, you could look for "violations to the 5S" method [5]:
    *   Sort: what is not essential or useful? What could be discarded?
    *   Straighten: what could not be rearranged in a more meaningful way?
    *   Shine: what is not clean?
    *   Standardize: are there random or subjective activities which could benefit from a standardized procedure?
    *   Sustain: what is already sorted, straighten, clean, and standardized, but not in a sustainable way?
7. In international study you may face a "language problem" when you cannot find a language with which both you and customers are comfortable. This

problem can be solved by involving local dealers, who have an interest to have better products in the future [3].

8. Experience shows that bias can be reduced if do not visit customers alone, but at least one colleague joins you. Ideally, one member should be from a customer-facing role such as sales or marketing, and one member should be from a technology- or operations-facing role such as engineering or production. This pairing minimizes the inherent bias of only one set of eyes and ears, and the differing perspectives enriches interpretation of the customer voice and behavior. If it is not possible to have two members on site, at least include others when reviewing the recordings. The challenge of bias reduction is discussed in more depth in Sec. 12.

4.8   TRANSLATE THE VOICE OF CUSTOMERS INTO CUSTOMER NEEDS

The "voice of customers", what you hear during visits and interviews, is usually about product features: what the product should do, look like, etc. Customers have the same tendency of your colleagues and of most humans: to talk about what they desire. However, although it may be tempting to follow the easy way and implement these features in a product, you should not. Customers think these features will solve their problems and create a benefit for them, but this assumption may be wrong; or the underlying need may be not a very important one; or they may not be willing to pay the feature price; or simply other features could solve the same pain in a more efficient way. Further, features indicated by customers are based on what they know today, which is yesterday's technology. New product development is about developing tomorrow's technology.

The voice of customers must be then understood (considering customer's context and emotions), clarified (only the customer can tell you if your understanding is right) and finally translated into needs. Typically, this can be achieved by asking "why?" multiple times, until, as recommended by Dr. Yoij Akao, one of QFD's founders, the need motivating a feature request is revealed ([6], p. 337).

Customer needs play a central role in this customer-centric method; therefore, it is important to agree on a standard format to formulate them. We define a customer need as:

- A single-issue statement, positively stated. It is often tempting to create complex statements with and/or conjunctions, negations, and subclauses, but it would be just a shortcut to save yourself the efforts necessary to achieve a complete clarification of customer needs.
- Describing a benefit to the customer: either, a pain solved, an opportunity enabled, an image enhancement.
- Independent of your product/service, technology, or features.

Adhering to these simple rules will ensure that your findings are unambiguous and suitable to be translated into a product design.

TABLE 4.5
Example of translation of customer voice into customer needs

Customer segment	Observation, Verbatims, and Clarified Items	Customer Need
Regular Joes	"When I drive by, I want people to think Armageddon."	I need to be noticed.
	"I want a big car."	I want other drivers to get out of my way.
	"What I like neve goes out of style! We need more chrome, and turf carpeting."	I need to appear classy.
	"I need a cup holder for my 2-liter water bottle!"	I need to easily grab my large cup.
	"I need a searchlight attached to my car so I can find it in a parking lot at night. Every car should have that!"	I don't have to remember where I parked.
	"I need a horn button I can smash when I'm mad! And it should sound like [a] gun fire"!"	I need to express my anger at other drivers.
		I need to appear classy.
	"My kids are in the backseat...driving me nuts."	My kids won't distract my driving.
		My kids won't interrupt listening to my music.

It is very important to document the connection between the voice of the customers and their needs so that this relationship can be reviewed and refined, or reevaluated, if needed. This can be conveniently done with the so-called "customer voice table".

As an example, the customer voice table for regular Joe's case study, based on the data in Table 4.4, is shown in Table 4.5. This analysis was not performed by the engineers in the mentioned Simpons episode who, pressed by the CEO, make the mistake of directly implementing Homer's requests without understanding the underlying needs, thus realizing an ultra-expensive "monstrosity" which ruins the company. Consider, for instance, the statement about the searchlight: it is clearly absurd, since if all cars had it, they would have been indistinguishable one from another. However, this request reveals an important need: to remember where your car is, which is often very difficult in big and crowded parking lots. Once the need is understood, if it turns out to be one of the most important, it is the role of designers to find a proper technical solution. How this can be achieved will be discussed in Chapters 5–8.

4.9   SUMMARY

A summary of the procedure outlined in this section is given in the example of Figure 4.6. Here the first three customers from "segment A" have been visited or interviewed. The feedback from the first customers has been collected using standardized questions. For the subsequent customers a CPM was available; thus, the feedback from them was directly annotated on a CPM sheet (one for each customer).

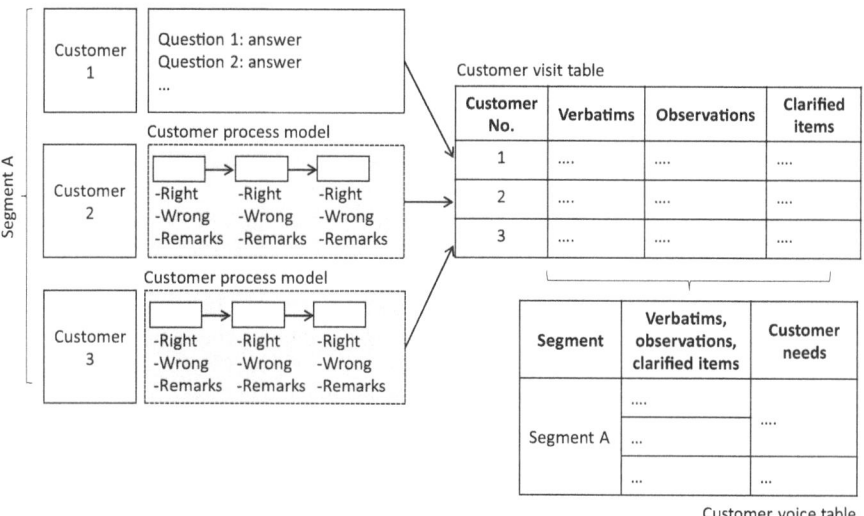

FIGURE 4.6   Summary of the procedures outlined in Chapter 4.

The most important data from visits are collected in the customer visit table, where one row is dedicated to each customer.

The most important data from the segment are collected in the customer voice table, which is used to derive customer needs from verbatims, observations, and clarified items.

NOTES

1   The common and ambiguous expression "steep learning curve" may indicate both situation of Figure 4.2 left (the curve is steep close to the origin, so you will learn a lot with few effort) and middle (it is hard to learn, like climbing a hill).

2   This behavior is sometimes indicated as "Pareto principle". Actually, Pareto (end 19th century) showed that the 80% of Italian wealth was owned by 20% of people. Today the term "Pareto principle" usually implies that most things in life are not distributed evenly [9]. In our case, a certain time dedicated to data collection at the beginning of the investigation is much more productive than the same time invested later.

3   Some team members may find the concept of "firing their mirror neurons" more stimulating than "empathy" and you can use it if it helps (although the relationship between mirror neurons and empathy is a subject of neuroscience that is beyond this discussion).

BIBLIOGRAPHY

[1]   Donaldson, K. M., Ishii, K., Sheppart, S. D., "Customer value chain analysis", Research in Engineering Design, vol. 16, pp. 174–183, 2006.

[2]   Hennink, M., Kaiser B. N. "Sample sizes for saturation in qualitative research: A systematic review of empirical tests", Social Science & Medicine, vol. 292, p. 114523, 2022.

[3] Mazur Glenn, H., Bylund, N., "Globalizing Gemba visits for multinationals", Transactions from the 15th Symposium on Quality Function Deployment, Monterrey (Mexico), 21–23 October 2009, p. 61.

[4] Vongpatanasin, T., Mazur, G., "Thai Brewery deploys QFD tools to tap into consumer motivation", ASQ, The Global Voice of Quality, December 18, 2012.

[5] Harkins, R., "Gimme Five: Count on the 5S improvement method in the lab", Quality Progress, August 2009.

[6] "QFD: The customer-driven approach to quality planning and deployment", Mizuno, S. and Akao, Y. editors, Asian Productivity Organization, Tokio, p. 337, 1994.

[7] Groening, M., Brooks, J. L., Simon, S. (Writers), Archer W. (Director), aired 1991-02.21, "Oh brother, where art thou?", season 2, episode 15, Adamson L. J., Brooks J. L., Groening M., Kogen J. (Producers), "The Simpsons", Gracie Films, 20th Century Fox Television.

[8] Jesso-White, J., Mazur, G. H. "DREAM/QFD to redesign new physician orientation at Rutland Regional Medical Center", Transactions from International Symposium on QFD, Portland (OR), 85–97, 2010.

[9] Juran J. M. "The economics of quality", in J. M. Juran, editor, Quality control handbook, McGraw-Hill, 1951.

5 Structuring Customer Needs

5.1 WHY STRUCTURE CUSTOMER NEEDS?

At this point in a market research, you typically have a list of many customer needs. You may have hoped to discover a few needs common to many customers, but this is rarely the case. It is not uncommon to have 20–30 needs at this point, which can be general as well as very specific. Some could be translated into expensive requirements; others could be relatively easy to implement. Few needs may have been disclosed by more than one customer, others by one customer only. Usually, the only certainties at this stage are the following:

- You cannot satisfy all needs with reasonable efforts.
- The needs are too varied to provide a clear description of customer situation.
- You cannot be sure that your list of needs is complete. On the contrary, it is likely that something is missing, which could be particularly important. No matter how many needs you get, they are always an incomplete set.

The goals of the structuring procedure described in this section are to solve these difficulties by understanding how customers think about their needs, identify a general framework in which all needs find a place, and find the missing needs. Only at that point you will be able to carry out a meaningful and fair comparison between comparable needs and prioritize them, which is the subject of the next chapter.

5.2 INTRODUCTION TO AFFINITY AND HIERARCHY DIAGRAMS

The affinity diagram is a simple and effective way to organize a set of unstructured elements. To create an affinity diagram means to arrange many elements in a chart by splitting them into groups. Typically, one creates a sort of pyramid with parent-groups which contain child-groups and assign headers to all of them.

This method was popularized in the 1960s by Japanese ethnologist Kawakita Jiro, who used it to integrate heterogeneous data that had been produced by his research [1]. Since then, the creation of this chart is also known as "KJ Method®".

It is important to realize that the affinity diagram reflects the mindset of the person who created it. Imagine, for example, that you want to know how people

DOI: 10.1201/9781003544845-5

think about animals, in particular, grizzly, dog, cat, pig, sheep, wolf, mountain lion, rabbit and cock. A convenient way to perform the classification is to write the names of the elements on cards and arrange them on a table or on a wall. Thus, you start by writing down the name of these animals on white cards, and you ask a farmer to arrange the cards in groups, and to create group headers whenever needed. They, who receive light gray cards to write down headers, could create a diagram like the one in Figure 5.1 (top).

The diagram contains much more information than the initial nine cards. It tells us that a farmer sees animals either as "domesticated" (which belong to them) or as "wild" (which are free) ones. The domesticated can be "pets", which provide help or company, or "farm" animals, which are an economic resource. Similarly wild animals can be "carnivores", a danger for livestock, or "herbivore", which the farmer can hunt as a source of meat or to prevent the destruction of their cash crops.

If you ask someone else to arrange the same cards, you are likely to see a different result. A student who is starting to receive classes of biology could create a diagram like the one in Figure 5.1 (bottom), which reflects a primitive zoologic taxonomy: now, for instance, the cat is in the same group of the mountain lion, as dog and bear do, due to their biological similarity.

An expert zoologist would surely argue that the diagram made by the student is naïve and scientifically incorrect. However, in our context, an affinity diagram is never wrong, since its goal is to reveal the way in which its creator thinks.

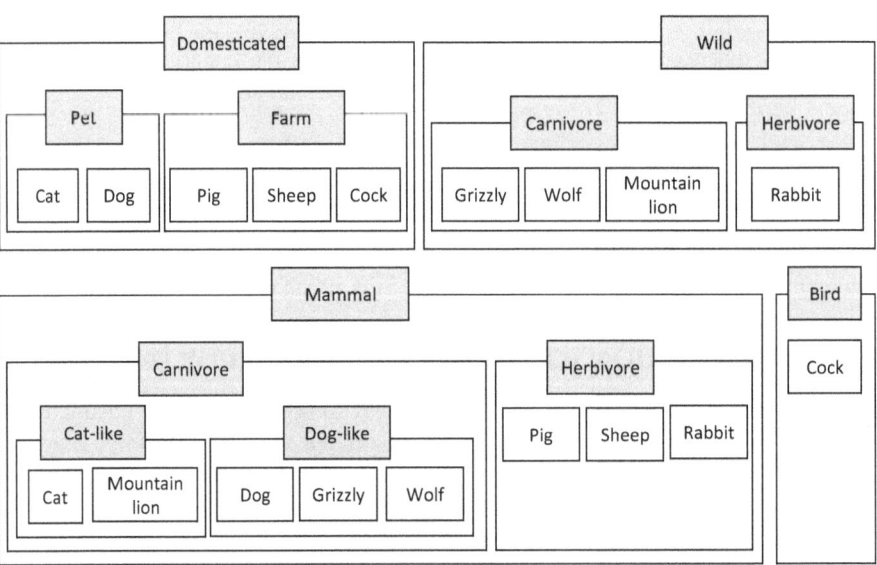

FIGURE 5.1 Example of different affinity diagrams which could be created by a farmer (top) or by a biology student (bottom) using the same nine cards (white). Light gray cards have been added as group headers during the creation of affinity diagrams.

Let us focus on the farmer example. Now that you have learned how they think about animals in general, it is worth trying to learn as much as possible from them by asking:

- To adjust the level of abstraction of each group. Are the members of each group comparable? Or are some of them more general or specific than others?
- Are there missing members? Are the members of the subgroup exhaustive? Special attention must be paid to groups with a single member: either the member and the group header are synonyms, or something is missing. In general, the members of any group should follow the MECE principle, that is, they must be "Mutually Exclusive" and "Collectively Exhaustive" [2]. Note that looking for missing needs is the most efficient means to uncover unspoken needs. To discover them you do not need an exceedingly high number of interviews and resources, but just enough to disclose the overall "architecture" of the hierarchy diagram.

Note that the analytic diagram is customarily drawn top to bottom. In QFD practice, in order to adjust the levels of abstraction and identify missing members, the diagram is usually rotated by 90° counterclockwise in such a way that the most general items are on the left and the most detailed ones are on the right. After this rotation the tree takes the name of "hierarchy diagram". The habit to use these orientations traces back to the introduction of the analytic and hierarchy diagrams (also called tree diagram) as part of the "7 new quality control tools" ([3], [4]), and it will be followed in this book too. Most people are comfortable with these conventions. However, different orientations may work better depending on the culture and the tools used to build the diagrams.

In our imaginary example, the farmer adjusts the diagram as in Figure 5.2. The elements they add to make the diagram exhaustive are indicated as dark gray boxes.

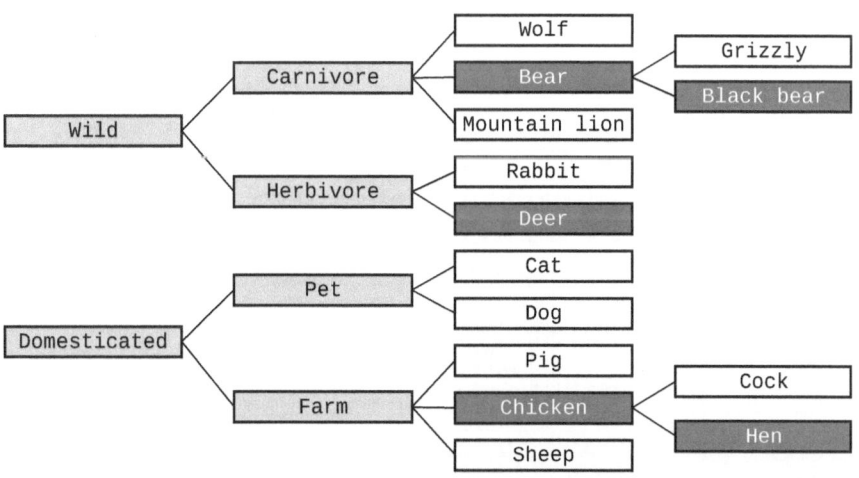

FIGURE 5.2 The farmer hierarchy diagram after having adjusted the levels of abstraction and having added the missing members (dark gray card).

The farmer noticed that one cannot compare pig and sheep (which are species of animals and can be either male or female), with a cock (which has a specific gender). He solved this inconsistency by adding "chicken" beside "pig" and "sheep" and putting the "cock" in the "chicken" group together with a "hen". He also decided that "wolves" and "mountain lions" compare better with a "bear": the grizzly is a specific bear, and other animals are bears, although different from the grizzly, like the "black bear". Finally, since the rabbit could not have been the only wild herbivore, he added another one he knows well: the deer.

5.3   AFFINITY AND HIERARCHY DIAGRAMS IN QFD

In the QFD method, you ask customers to create an affinity diagram to structure their needs. Many examples of affinity diagrams from real studies can be found in the literature ([5]–[7]).

To explain in detail how the creation of an affinity diagram with customer needs works, we will use the needs of "regular Joe" example from Table 4.5. Let us assume that these needs come from many customer interviews and visits to parking lots and workshops.

What you must do first is to write the needs on cards and ask a group of customers, from the top segment, to create the affinity diagram. Practical hints to organize and drive their work are given at the end of this chapter. Imagine they will produce the diagram of Figure 5.3, which provides a lot of insight into the mindset of these customers. At the primary level, we see they would buy a car primarily to improve their image, and that they need to be relaxed when using it (probably because their present car does not provide satisfactory user experience). Note also that needs at primary and secondary levels are too general to drive product development. On the other hand, items at tertiary level are needs for which your engineers could find a solution.

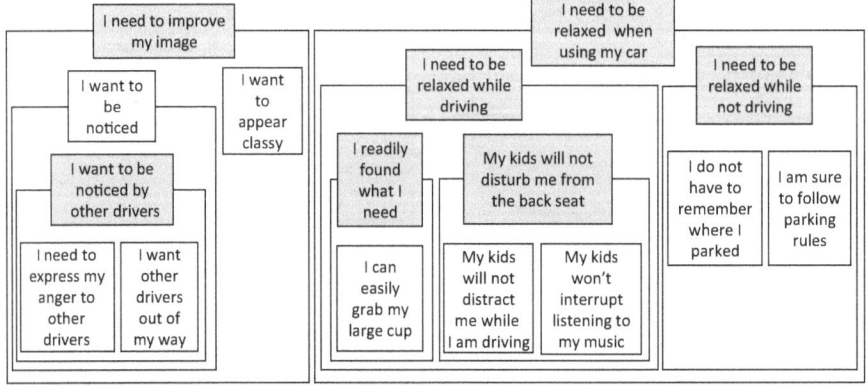

FIGURE 5.3   Example of affinity diagram created from the "regular Joe" needs of Table 4.5. White cards: customer needs. Light gray cards: headers created by customers at the time of building the diagram.

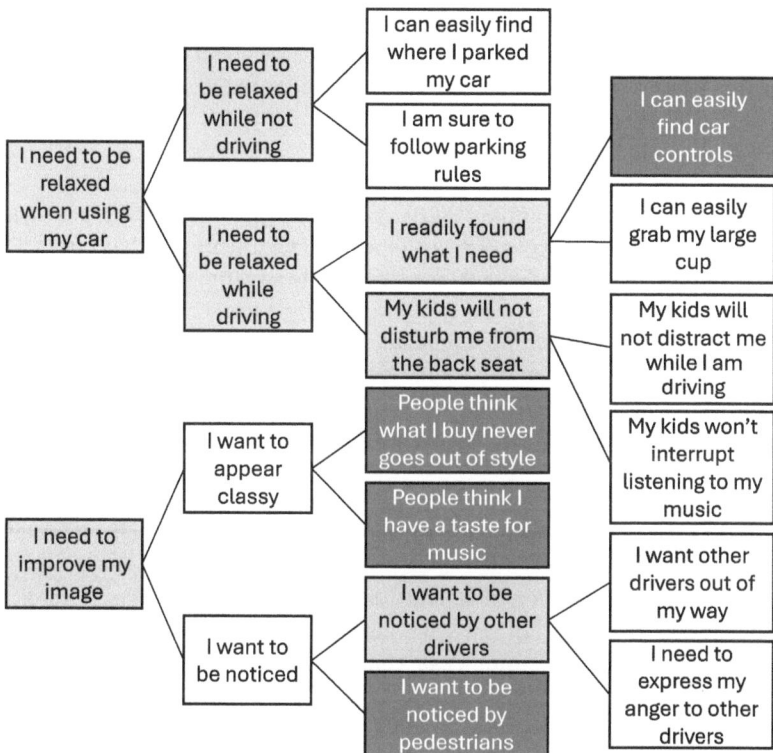

FIGURE 5.4   The diagram of the previous example after being adjusted and completed. White cards: initial needs; light gray cards: items added as group headers; dark gray cards: items added to complete the diagram.

Additionally, these solutions would probably have a comparable complexity so that they will make sense as alternatives.

To be usable for needs prioritization, this diagram still needs few adjustments:

1. The secondary level need "I want to appear classy" is not detailed at tertiary level. This is unsatisfactory, since we have seen that tertiary needs are extremely useful for product development.
2. The needs "I want to be noticed" and "I readily found what I need" are headers of groups with only one element, and you shall suspect that there are more needs to be discovered within these groups.

When you ask the team of customers to solve problems (1) and (2), they could generate the hierarchy diagram in Figure 5.4. Now you have at least tertiary needs for all branches, and every primary and secondary need has at least two children. This is a diagram which would be suitable for needs prioritization.

Note that all customers' requests have been reduced to two basic needs: improve their image and be relaxed. These needs would have fit very well with Homer

Simpson's personality too: he does not need mileage or low fuel consumption since he lives and works in his hometown. He does not care about performance either; the powerful engine is merely intended to rev up and make noise.

5.4 PRACTICAL PROCEDURE

The best way to create the affinity and hierarchy diagrams is to put together a small group of customers from your top customer segment. This will help to cancel out individual fluctuations in judgment, and to generate an affinity diagram that is representative of the segment. If you are exploring multiple top segments at the same time, customers should be grouped by segment and not co-mingled.

It is important to prepare everything you may need in advance. You must efficiently organize the customers' activity so that you maximize the information you can collect during the time they can dedicate to you. You need:

- Explanatory cards to teach the method. You can use animals, as we did above, or the items you think are more appropriate. All cards must be readable at arm's length.
- Customer needs written on another set of cards. Simply write on cards all the needs discovered during visits and interviews. It does not matter if these needs were specific to one customer or common to many. From now on, customers must be confronted with all needs.
- Be sure that a table is available to distribute the cards, or a board to hang them.

Before starting with the affinity diagram, you must clearly explain its construction rules with the explanatory cards. Give the group a role (if you use "animals", you can choose between farmer, chef, zoologist, etc.). If you have enough customers, split them into two or more teams with different roles so that they will understand the importance of the role. In this case you will need multiple packs of cards and multiple tables or boards.

Ask the team (or the teams) to arrange the cards into groups, according to the point of view of their role. They should accomplish this task without talking. This will ensure a quick convergence without extenuating discussions and attempts of one customer to change the mind of others with verbal argumentation. Only when cards have been sorted that team members can start talking to create headers. Clarify that every group of cards needs a header: customers can promote an existing card to header status or create a new one. At this stage, groups can be combined into super-groups too.

When the affinity diagram is completed, customers can rotate it 90° counterclockwise and start creating the hierarchy diagram. Ask them to adjust the level of abstraction in every branch. If necessary, they can move cards and create new entries to implement the MECE principle. Invite them to pay special attention to groups with a single element: either it is a synonym of the header or it has at least one sibling at the same level of abstraction. Finally, review the results with the customers and answer possible questions. If there are few groups, it is worth discussing the difference between the diagrams they created, and the relationship between these differences and group roles.

When the explanatory exercise is completed, distribute the cards with customer needs and invite the customers to repeat the procedure you taught them. Caution them that any new card they fill out, a header or a need necessary to satisfy the MECE principle, must be a "customer need" according to the definition given in Section 4.8: a simple statement which expresses the benefit to them of their problem solved, their opportunity enabled, or their image enhanced. They should avoid writing on cards technical solutions and product features. It is also important to remind the customers that needs should express a single concept, and therefore they should avoid terms like "or", "and", *and* "etc."

BIBLIOGRAPHY

[1] Scupin, R., "The KJ method: A technique for analyzing data derived from Japanese ethnology", Human Organization, vol. 56, June 1, 1997.

[2] Minto, B., The pyramid principle, logic in writing and thinking, Pearson Education, 2021.

[3] Mizuno, S., Management for quality improvement: The 7 new QC tools, Taylor & Francis, 1988.

[4] Brassard, M., "The memory jogger plus +", GOAL/QPC, 1996.

[5] Jesso-White, J., Mazur, G. H., "DREAM/QFD to redesign new physician orientation at Rutland Regional Medical Center", Transactions from international symposium on QFD, Portland (OR), pp. 85–97, 2010.

[6] Vongpatanasin, T., Mazur, G., "Thai Brewery deploys QFD tools to tap into consumer motivation", ASQ, The Global Voice of Quality, December 18, 2012.

[7] Mazur Glenn, H., Bylund, N., "Globalizing Gemba visits for multinationals", Transactions from the 15th symposium on quality function deployment, Monterrey (Mexico), p. 61, 21–23 October 2009.

6 Prioritization of Customer Needs

At this point you have a set of customer needs, relatively complete and organized in the "big picture" provided by the hierarchy diagram. You have gained a general understanding of the context in which customers live or work, and why they have such needs. The next step is to prioritize these needs together with your customers. This section deals with basic prioritization concepts, the general procedure, and a few simple methods. The next section is devoted to a most rigorous prioritization method: the pairwise comparison borrowed from the Analytic Hierarchy Process.

It is essential to perform the need prioritization with many customers, at least with all those who were interviewed or met in Gemba visits. By combining their judgments, you remove individual fluctuations (i.e. evaluations different from the customer segment average which are due to the personality or due to the very specific situation of different customers) and disclose the real top needs of the customer segment.

Prioritization involves three fundamental steps:

1. Collect the priorities of individual customers. In this chapter we assume that prioritization is carried out with one customer at a time. The possibility to gather many customers and carry out a panel-based prioritization is discussed in Section 11.1.
2. Combine individual priorities to get the mean priority of every need. At this point you can identify top customer needs.
3. Verify the homogeneity of the customer segment. This is a double check that the customers in the segment share the same top needs.

6.1 COLLECT PRIORITIES FROM INDIVIDUAL CUSTOMERS

During the prioritization of customer needs, it is important that the customer can constantly see the full hierarchy diagram. Therefore, if you cannot meet the customers in person, use a video call. This will generate more consistent results and will minimize the possibility of misunderstandings. You should avoid to collect scores with a questionnaire, since the risk that the customer misunderstood the whole procedure would be unacceptable.

 DOI: 10.1201/9781003544845-6

There are many possibilities to evaluate the relative importance of needs with a customer. However, the first step is always an introduction during which you present the customer the hierarchy diagram, review the needs with them, and explicitly ask if they understand their meaning. If necessary, you will have to provide additional explanations, and then only you can proceed with the prioritization.

You can invite the customer to use your favorite prioritization method but be aware that the customer may want to use a different one. Especially if you try to perform quantitative prioritization with scores, it is not uncommon to meet customers who want to skip the procedure saying, "I know which is my top need". It is not a good idea to argue with customers because in the end you want to understand them without introducing biases. In such a situation, it is wiser to give in: it is not critical that different customers provide their priorities with different methods, you will just need some extra care during data analysis to combine them.

6.1.1   QUALITATIVE PRIORITIZATION

The simplest and faster method to prioritize the needs is as follows:

- Directly ask the customer which need is for them the most important to satisfy and place a mark in the diagram.
- Inquire why this need is so important for them and ask confirmation that they would be willing to pay for a product satisfying this need.
- Optionally, you may even inquire how much money they would pay to satisfy this need. This would provide an indication of the magnitude of need importance.

It is not rare that a customer has difficulties in defining the top need: maybe there are two needs with close urgency to be satisfied, or they may feel that some other need, even being not the top one, should not be completely neglected. If the customer picks up more than one need in the diagram, do not worry: this is again something you can take care of during data analysis. It is good practice to ask them if the needs they indicated are equally important to be satisfied, or if one stands above the other, and take note of this information.

6.1.2   QUANTITATIVE PRIORITIZATION

The easiest way to carry out a quantitative prioritization of needs is as follows:

- Focus on a group of needs, starting from the primary level grouping of the hierarchy diagram (which is typically the very general group at the left).
- Ask the customer to assign a numerical score to every need in the group. Clearly explain to the customer that the scores should reflect the relative priority of the needs in the group. You will normalize the scores later; thus, it does not matter if they sum up to a specific value (e.g. 1, 10, and 100%). For instance, if you have four needs in a group, and if the customer feels that the first is three times

more important than any other, he may score them (3, 1, 1, 1), or (1, 1/3, 1/3, 1/3), or (45%, 15%, 15%, 15%), etc. Any of these sets is acceptable for you, since they have the same meaning; just let the customer feel comfortable with his evaluation. You may be tempted to ask the customer to split 100 points between the needs in a group, but this is dangerous: it works fine only if the group has two members; in any other case, the customer will waste mental efforts thinking about "how can I communicate my priorities and still have numbers summing up to 100".

• Move your focus to the child group of the need with highest priority and repeat the procedure till the right-most level of the diagram is reached. Doing so you will identify a "highest priority" path through the hierarchy diagram.

• If necessary, catch up with the prioritization of lower-priority branches. You should do it only if necessary, for instance, if you have two needs with the highest priority in the same group, or if a need has a priority just slightly smaller than the priority of the top one. This frugality helps to avoid wasting the customer's time and attention discussing needs which are not important for them or, even worse, run out of time before completing the evaluation.

An example is shown in Figure 6.1. This customer used percentage score at primary and secondary levels; at tertiary level, his reference was the priority of the least

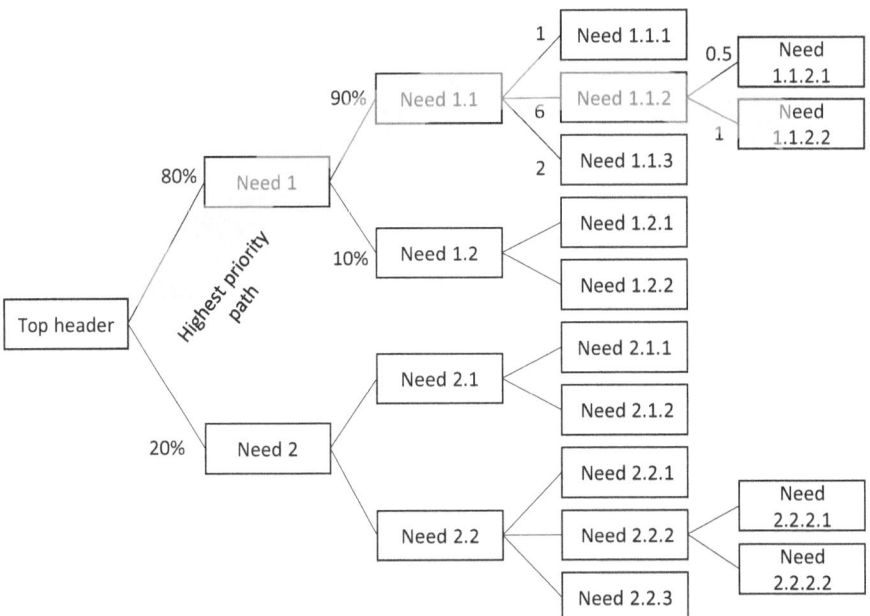

FIGURE 6.1 Example in which a customer provided a quantitative prioritization of the needs in a hierarchy diagram. Note that the customer provided scores with different normalizations in the groups, and that only the highest priority path was investigated (children of Need 2 and Need 1.2 were neglected). The highest priority path is shown in gray.

important need, which he set to 1; at quaternary level he used 1 for the priority of the most important need and assigned priorities to the other accordingly.

These quantitative scoring methods grant the customer the flexibility needed in complex scenarios in which more needs have similar importance. Their main drawback is that the numerical priorities you collect may poorly represent the need importance. Experience teaches us that needs prioritization follows the Weber–Fechner law: the need satisfaction importance is the stimulus for customer's mind, and the priority that they communicate to you is the perception. Thus, the relationship between the importance of satisfying a need and priority is sublinear: a substantial change in importance is needed to generate a small change in the priority. For instance, you have no guarantee that a need with 40% priority is really two times more important than a need with 20% priority, and that customers would pay double to have it satisfied. As explained in Chapter 7, this drawback can be solved by performing a pairwise quantitative comparison with natural language.

6.2 QUALITATIVE COMBINATION OF MULTIPLE JUDGMENTS

If you can identify a single top need for every customer, you can simply crate a flow map like the one in Figure 6.2. In this case, for every customer, you draw the "highest priority path" curve from the left of the diagram to the right. The curve must cross the needs which, at each level, exhibit the highest priority path. Drawing the map often clearly reveals both the top need of the segment (the one receiving more scores, which

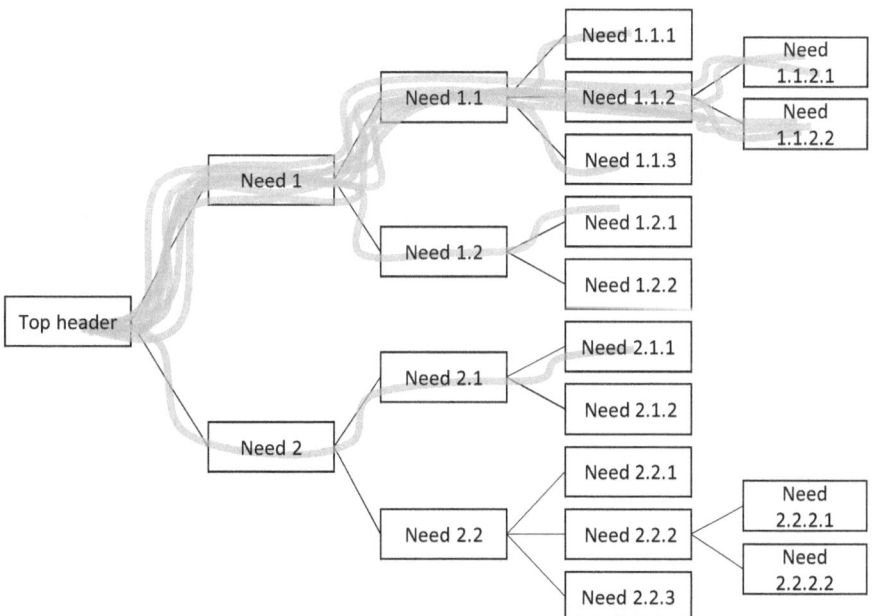

FIGURE 6.2 Example of judgments from nine customers belonging to a homogeneous population shown as a flow map.

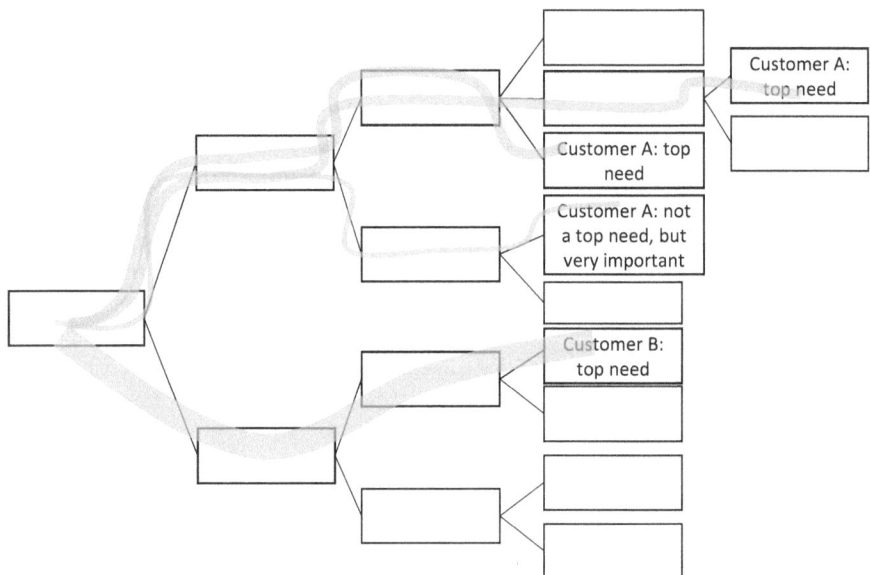

FIGURE 6.3 In this example the priority of two customers (A and B) is represented with a flow map. Customer B has one top need, which is reached with a path indicated by a thick curve. Customer A has two top needs and another very important need, which are reached with paths indicated by thinner curves.

statistically represents the mode) and the most important region of the diagram. In the figure the top need is "Need 1.1.2.2" but, besides solving it, you will probably want to improve all the product features related to the Need 1.1, which indicates an area needing improvements.

With this technique and some artwork skills you can also clearly describe complex situations in which you do not have one single top need for every customer. Consider, for instance, the example of Figure 6.3. Here we have two customers: customer B expressed a single top need, while customer A could not help having two top needs. Besides that, customer A stresses that another need, although not a top one, is for him very important. This has been considered by drawing lines with different thickness. Customer A is represented by three curves, two of which (top needs) have double the thickness of the third (non-top need), in agreement with Weber--Fechner law. The sum of all thickness of customer A curves (in the original artwork 6, 6, and 3 thickness points, respectively) equals the thickness of the single curve used for customer B (15 points).

This method is very intuitive and generates good-looking charts your colleagues will probably love. In Figure 6.4 you can see a real-life example from industry. The hierarchy diagram topology is exactly the original one, although the needs description has been removed because of confidentiality reason. You can appreciate how three high-priority regions, which can be addressed during design, are clearly highlighted.

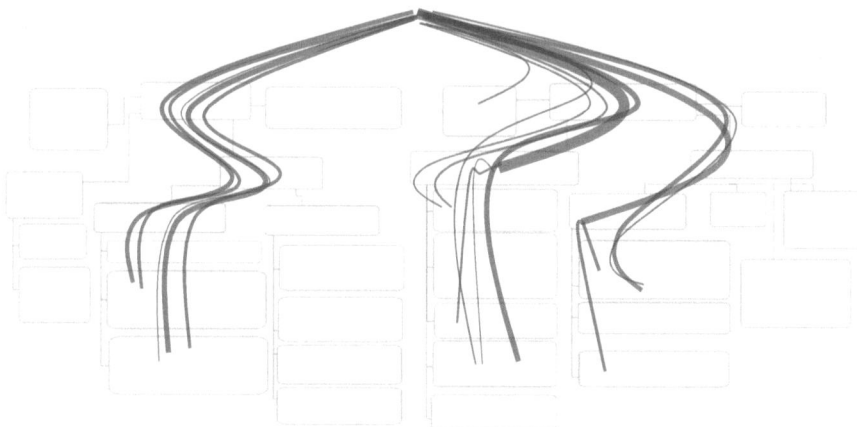

FIGURE 6.4 Example of qualitative prioritization from a real project in manufacturing. This diagram refers to an intermediate project stage and includes the prioritization provided by five customers. The flow map artwork was created with MS PowerPoint.

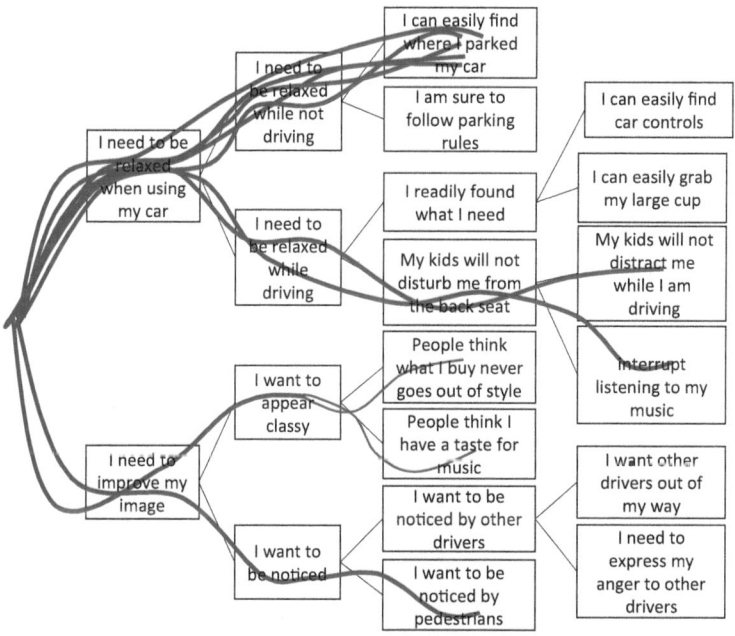

FIGURE 6.5 A possible qualitative prioritization of needs which could have been obtained by prioritizing the hierarchy diagram of regular Joe's case study with ten customers.

As a further example, a possible flow map which could have been obtained in regular Joe's case study by interviewing ten customers is shown in Figure 6.5. In this case all customers but one indicated a single top need.

6.3   QUANTITATIVE COMBINATION OF MULTIPLE JUDGMENTS

If your prioritization consists of numerical priorities, the first step is to normalize the priorities p_{local} within every group of needs so that their sum is 1. An example is given in Figure 6.6. At primary level the customer gave the priorities (20%, 80%): they are already normalized to 100%, thus you merely need to rescale their sum to 1: (0.2, 0.8). At tertiary level, the customer gave the priorities (1, 6, 2) to a group of three needs. With normalization you get:

$$\frac{(1,\ 6,\ 2)}{(1+6+2)} = (0.11,\ 0.67,\ 0.22). \tag{6.1}$$

Subsequently, for every need, you calculate the average local priority as the arithmetic mean of local priorities assigned by different customers. An example with two customers is given in Figure 6.7. Note that at quaternary level customer no. 2 did not provide a scoring because these needs were not the most important to him, thus he is not considered in the mean calculation.

Optionally, for every need you can calculate the global priorities. For the primary level local and global priorities are the same. Starting from secondary level of the diagram, the global priority of a need is obtained by multiplying its local priority by the global priority of its parent. The normalized local priorities and the global

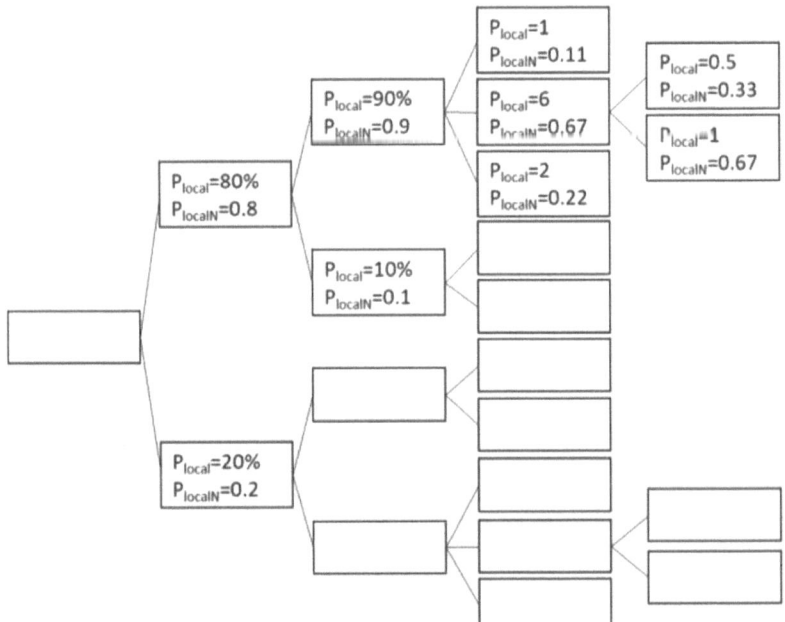

FIGURE 6.6   Example of quantitative prioritization. P_{local}: priority given by the customer. P_{localN}: the same of P_{local} but normalized to 1 within a group.

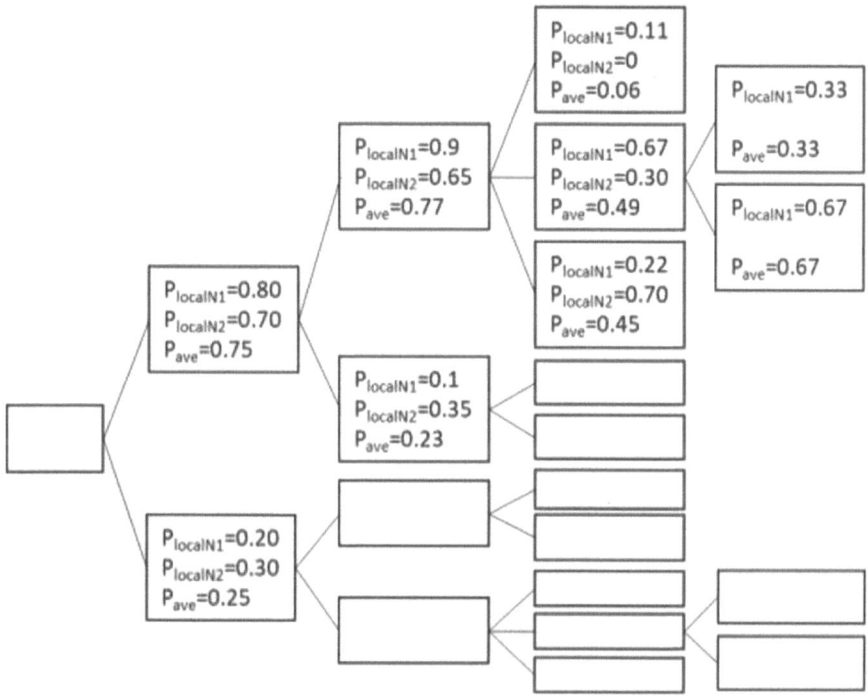

FIGURE 6.7 Example of averaging the local priorities given by two customers (1 and 2). $P_{localN1}$ and $P_{localN2}$ are the normalized local priorities provided by the customers and P_{ave} is their arithmetic mean.

priorities for the example of Figure 6.7 are shown in Figure 6.8. Note that within a group the sum of global priorities is the global priority of their parent, whilst the sum of normalized local priorities is 1; these two facts can be used to discover typing and calculation errors.

Global priorities will be necessary if you want to perform a quantitative prioritization of product functions (as explained in Section 8.5). Besides that, they can help you to better understand the hierarchy diagram. In the example in Figure 6.8, the highest priority path develops through Need 1 and Need 1.1. At this point you have different alternatives, since Need 1.1.2 and Need 1.1.3 have similar priorities, and Need 1.1.2 has two sub-needs:

- If you have enough R&D resources, you can try to address Need 1.1.2 and Need 1.1.3 completely. This will ensure you will deliver high value to customers.
- If you do not have enough resources, you can completely address Need 1.1.2 by solving Need 1.1.2.1 and Need 1.1.2.2.
- As an alternative to save even more resources, you can solve Need 1.1.3 only.

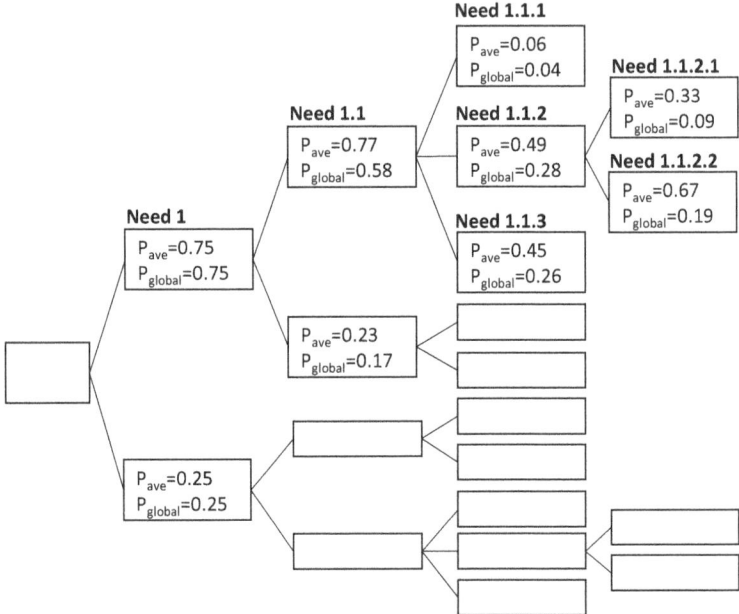

FIGURE 6.8   Relationship between local and global priorities.

Note that although Need 1.1.2.2 sits at the end of the highest priority path, it would be not devisable to solve it alone, since it has a lower global priority than Need 1.1.3.

A further example of qualitative prioritization based on regular Joe's case study is shown in Figure 6.9. Since not all the branches of the hierarchy diagram have needs at quaternary level, it is better to focus on the tertiary level to identify the top needs. Here you have a clear winner, "I can easily find where I parked my car": it sits at the end of highest priority path, and it has the highest global priority too (0.48). Another need you may want to consider is "my kids will not disturb me from the back seats" (global priority 0.24). Both these needs are about being relaxed while driving. If you want to consider image issues, the most important need in the bottom half of the diagram is "I want to be noticed by pedestrians" (global priority 0.12).

In order to calculate global priorities, we have first averaged local priorities from different customers. We will keep on using this habit throughout this book: this procedure is simply more practical because we always perform quantitative prioritization by comparing needs within the same group in the hierarchy diagram. You could have however followed a different procedure by (1) calculating first the global priorities for every customer and (2) averaging them. In general, these two methods provide different results, and none of them is correct or wrong: they simply reflect two different prioritization philosophies. Luckily, significant discrepancies can arise only if you have too few customers, or if the customer segment is not homogeneous. If you are interested, this subject is discussed in more detail in Appendix A.

Needs Primary level	Priority Local	Global	Needs Secondary level	Priority Local	Global	Needs Tertiary level	Priority Local	Global	Needs Quaternary level	Priority Local	Global
I need to be relaxed when using my car	0.72	0.72	I need to be relaxed while not driving	0.67	0.48	I can easily find where I parked my car	1.00	0.48			
						I am sure to follow parking rules	0.00	0.00			
			I need to be relaxed while driving	0.33	0.24	I readily found what I need	0.00	0.00	I can easily find car controls		
									I can easily grab my large cup		
						My kids will not disturb me from the back seat	1.00	0.24	My kids will not distract me while I am driving	0.50	0.12
									My kinds won't interrupt listening to my music	0.50	0.12
I need to improve my image	0.28	0.28	I want to appear classy	0.45	0.13	People think what I buy never goes out of style	0.50	0.06			
						People think I have a tast for music	0.50	0.06			
			I want to be noticed	0.55	0.15	I want to be noticed by other drivers	0.20	0.03	I want other drivers out of my way	0.30	0.01
									I need to express my anger to other drivers	0.70	0.02
						I want to be noticed by pedestrians	0.80	0.12			

FIGURE 6.9 Imaginary quantitative prioritization of customer needs for regular Joe's case study.

Note: Numerical values are such that the same results of the qualitative prioritization of Figure 6.5 are obtained.

6.4   SWITCH BETWEEN QUALITATIVE AND QUANTITATIVE PRIORITIZATION

It is possible that a part of customers provide qualitative priority, and others quantitative priorities. To combine them, you need to transform qualitative into quantitative, or vice versa.

The conversion of quantitative priorities into qualitative ones poses no problems: just draw a curve through the highest priority path and you will get a flow map.

If you want to convert qualitative priorities into quantitative ones, you can follow the example of Figure 6.10. Here the customer qualitatively indicated two top needs and another one with remarkable importance, although not a top one. To make this judgment quantitative you can proceed as follows:

- Assign a global priority to the needs mentioned by the customer. In agreement with the stimulus-perception lay, the top needs should have a score about double the non-top need – for instance, 1,1 and 0.5.

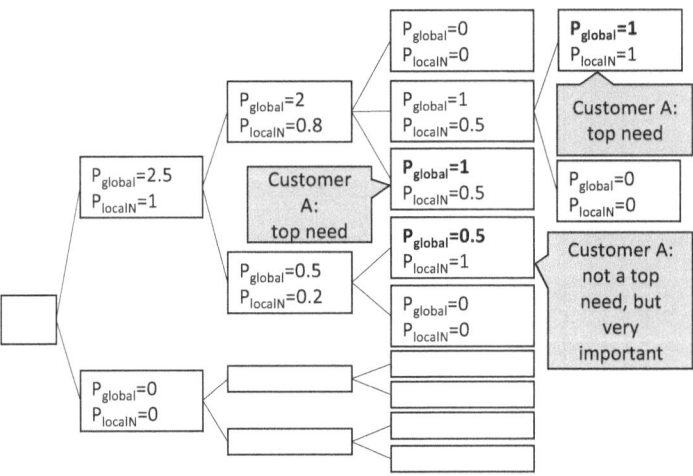

FIGURE 6.10   Converting a qualitative prioritization into a qualitative one.

- Assign a global priority 0 to the other needs in the same groups of mentioned needs.
- Moving from right to left in the hierarchy diagram, assign to parents the sum of global priorities of their children.
- Calculate the local priorities by normalizing the global one within every group.

6.5   VERIFY HOMOGENEITY OF JUDGMENTS

Up to now we have always assumed that the customers involved in prioritization are members of the most important customer segment and, therefore, they share the same top need. Now, it is the right time to verify our hypothesis. It would not be surprising that our initial understanding of customer segments was not very accurate, and that we have mixed up customers of sub-segments of which we ignored the existence. The homogeneity hypothesis can be verified by checking if the top needs of customers fall in the same area of the hierarchy diagram. It is not necessary that all customers have the same top need in the right end of the diagram, and few outliers are acceptable, but most customers should have some high priority need in a common region. If a part of customers favors a certain region of the diagram, and the rest another region, then the segment is not homogeneous, and customers should be split in two sub-segments before combining their judgments.

This verification can be easily performed by drawing the flow map. Even if you have collected percentage scores, you can always identify the top need of each customer and draw the flow map. As an example, a diagram like the one in Figure 6.2 tells you that your initial hypothesis is correct because most preferences are flowing in the same direction. Of course, you may have some outlier (Need 2.1) and some dispersion (Need 1.2), but there is a clear general trend.

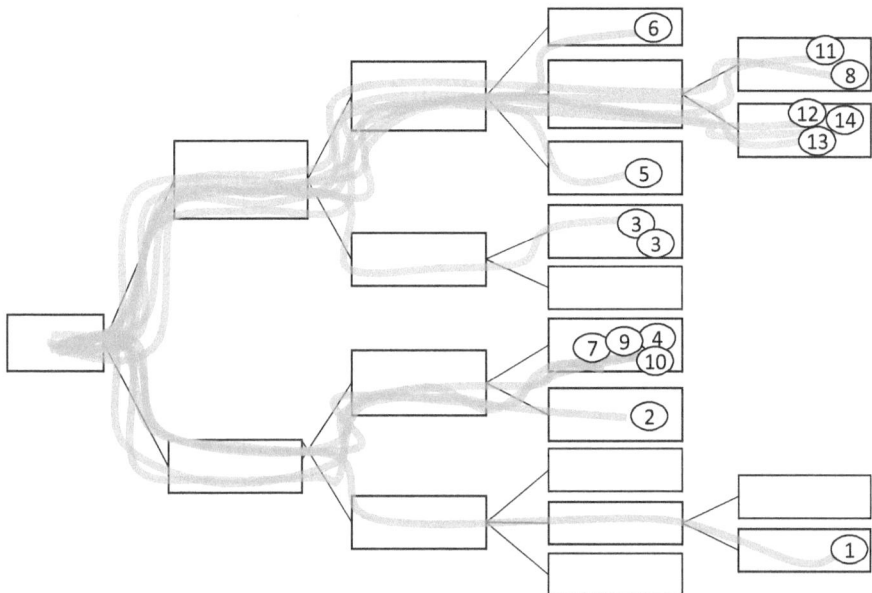

FIGURE 6.11 Example of judgments from 14 customers belonging to two different populations shown as a flow map.

On the other hand, a diagram like the one in Figure 6.11 indicates that your customers do not belong to a uniform population. Most customer preferences (8) point to the top branch and flow through Need 1, Need 1.1 and Need 1.1.2. There is however a substantial group of five customers with a completely opposite behavior. They favor the branch of Need 2, with a clear trend toward Need 2.1 and Need 2.1.1.

Note that a flow map with multiple "flow streams" is not sufficient to indicate an inhomogeneous population. Decisive is the fact that the curves flowing in different areas belong to different customers. For instance, the real-life example of Figure 6.4 belongs to a very uniform population, in which every customer has high priority needs in different areas of the diagram.

If the population is not homogeneous, it makes no sense to average the numerical priorities, because in this way you would simply hide what you have just discovered. This is shown in the example of Table 6.1, where a simple situation with three needs scored by ten customers is considered. The customers were thought to belong to the same segment, but they actually belong to two sub-segments with different priorities. Customers in sub-segment A exhibit a strong preference for Need 1 and Need 2, while customers in sub-segment B exhibit a strong preference for Need 3. If you divide the customers into two sub-segments and then perform the averages, you can see that for sub-segment A the top priority is Need 1, Need 2 is still important and Need 3 unimportant. For sub-segment B, Need 3 is the top one, Need 2 is still important, and Need 1 is unimportant. If on the contrary you average priorities over all customers,

TABLE 6.1
Example of data showing the effect of performing averages over a nonhomogeneous population

	Customers priorities												
	Sub-segment A (Customer no.)					Sub-segment B (Customer no.)					Average sub-segment A	Average subsegment B	Average of all customers
	1	2	3	4	5	6	7	8	9	10			
Need 1	0.7	0.4	0.6	0.5	0.8	0	0.2	0	0	0.1	0.60	0.06	0.33
Need 2	0.1	0.6	0.3	0.5	0.2	0.2	0.2	0.5	0.6	0.2	0.34	0.34	0.34
Need 3	0.2	0	0.1	0	0	0.8	0.6	0.5	0.4	0.7	0.06	0.60	0.33

Note: For every customer the cell with the top priority is highlighted in dark gray. Cells with priorities still important but not "top" are highlighted with light gray.

you get that the three needs have the same priority, and thus you do not gain any understanding about customer needs.

In general, if the customer population is not uniform, the correct procedure you should follow is as follows:

1. Understand which needs of the customers make them a member of one of the different sub-segments.
2. Identify which of the segments you have just discovered is the most important for you. This will require reflections about your business. Remember that the number of customers from sub-segments in your sample is not necessarily representative of sub-segment size.
3. Remove the customers who do not belong to the most important sub-segment and enroll new ones to replace them, to ensure a good statistic.
4. In the end, double-check again that the new group of customers is homogeneous.

7 Pairwise Comparison

7.1 INTRODUCTION TO PAIRWISE COMPARISON

This section is devoted to a rigorous method suitable to quantitatively prioritize the needs within a group: the "pairwise comparison" [1]. Pairwise comparison is a tool of "analytic hierarchy process" (AHP), a consolidated method for multi-criteria optimization.

The typical steps of pairwise comparison are shown in Table 7.1. In the table, four needs (Needs 1–4), which belong to the same hierarchy diagram group, are compared to determine their local priority P. The pairwise comparison process consists of three steps:

1. You ask the customer to compare every item against all the others. For every pair: first, you ask which one of the two is the most important to improve, then how much. You offer the customer a list of possible answers based on natural language, like moderately more important, strongly more important, etc.
2. Using an empirical conversion scale, you convert the answers based on natural language into numbers, which represents the ratio of the priorities of the two elements.
3. Using the AHP calculation algorithm, which will be discussed in the next sections, you convert priority ratios into priorities. As a bonus, the calculation provides the "consistency ratio", which tells you if the answers given by the customer are consistent. An example of severe inconsistency would be:
   • Need 1 is more important than Need 2.
   • Need 2 is more important than Need 3.
   • Need 3 is more important than Need 1.

Before going into details, we will highlight the pros and cons of pairwise comparison so that you can better judge if the benefits provided by this method are worth the extra effort it requires in comparison with simpler prioritization procedures.

7.2 PROS AND CONS OF PAIRWISE COMPARISON

Pairwise comparison is an accurate tool to extract quantitative priorities from subjective judgments, potentially more accurate than all other methods reviewed up to

 DOI: 10.1201/9781003544845-7

TABLE 7.1

Example showing the main pairwise comparison steps

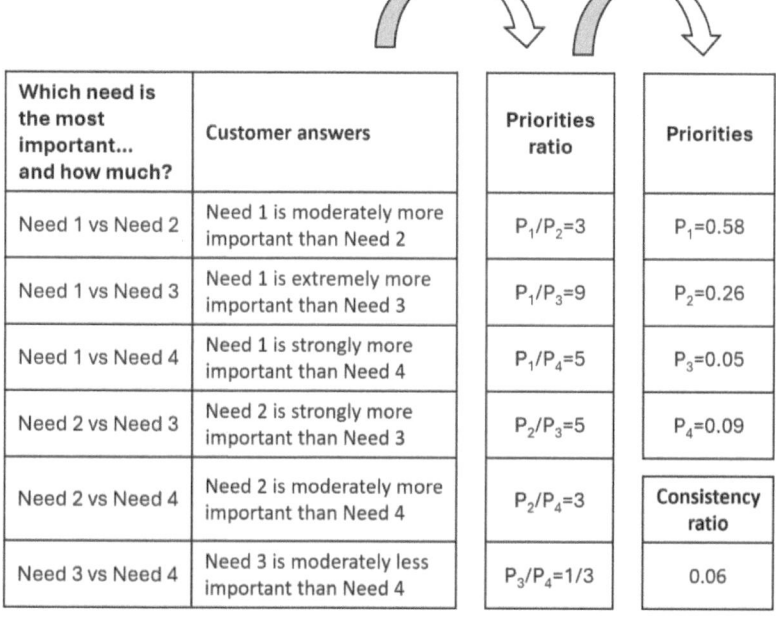

Empirical conversion scale   Algebra

Which need is the most important... and how much?	Customer answers	Priorities ratio	Priorities
Need 1 vs Need 2	Need 1 is moderately more important than Need 2	$P_1/P_2=3$	$P_1=0.58$
Need 1 vs Need 3	Need 1 is extremely more important than Need 3	$P_1/P_3=9$	$P_2=0.26$
Need 1 vs Need 4	Need 1 is strongly more important than Need 4	$P_1/P_4=5$	$P_3=0.05$
Need 2 vs Need 3	Need 2 is strongly more important than Need 3	$P_2/P_3=5$	$P_4=0.09$
Need 2 vs Need 4	Need 2 is moderately more important than Need 4	$P_2/P_4=3$	Consistency ratio
Need 3 vs Need 4	Need 3 is moderately less important than Need 4	$P_3/P_4=1/3$	0.06

now. However, this potential becomes reality at a price: it takes time to go through all the questions necessary to collect input data. With a group of N elements to compare, the number of questions is:

$$(N^2 - N)/2 = N(N - 1)/2. \tag{7.1}$$

This number grows as $\sim N^2$, as shown in Table 7.2. For a number of needs to be compared from 2 to 7, the number of questions to be stated is 2, 3, 6, 10, 15, and 21, respectively. In contrast, simply asking a priority estimate to customer just requires N questions.

More data collected with more questions means more information available to accurately estimate priorities, but also more efforts and time required to the customers. An interview which takes too long may let you run out of time, so that the evaluation is completed in a rush, or is not completed at all. Additionally, the interviewee can become bored, or lose focus, and could start providing unreliable answers. You may spot distracted answers by looking at the consistency ratio, but to correct them you will have to review the responses together with the customers, which will take even more time. Finally, not all the customers are comfortable with the pairwise comparison. Depending on their attitude, they could react in a negative way. A few may

TABLE 7.2
Number of pairwise comparisons necessary to prioritize
N elements [2]

No. of Elements to Prioritize (N)	No. of Comparisons ("Questions")	RI
2	1	(not applicable)
3	3	0.58
4	6	0.90
5	10	1.12
6	15	1.24
7	21	1.32

Note: The RI index, used later in this chapter, is indicated in the last column.

be annoyed by your request to put on the side the big picture and to compare pairs of needs instead. Others may even refuse to perform the pairwise comparison, ranking instead all the needs in a group at once.

The hierarchy diagram is of great help to save time since you can neglect the needs of low-priority branches. It is however important that groups do not contain too many elements, otherwise the number of paired questions could be overwhelming. Let's consider few examples:

- With three levels of groups, and groups of 3 elements, assuming you can follow a single branch, you will need to state $3 \cdot 3 = 9$ questions. This is a very reasonable effort.
- With four levels of groups, and groups of four elements, still assuming you can follow a single branch of the diagram, you need to state $4 \cdot 6 = 24$ questions, which is a nontrivial effort.

In conclusion, there is no general rule telling you whether you should use the pairwise comparison or a simpler alternative. It depends on your project resources (time, people, and budget), your customer segment, the size of your hierarchy diagram, the mindset and availability of your customers, and your confidence with mathematics. In the end you will have to decide for yourself.

7.3   QUANTITATIVE SCORING WITH NATURAL LANGUAGE

To circumvent the challenge created by the Weber–Fechner law, you can ask the customer to judge the importance of needs with terms from everyday language, and then translate these verbal judgments into numbers by using some empirical, tested scale. Saaty initially proposed the popular linear scale for priority ratios ranging from 1 to 9 [1, 2].

He successfully tested the scale with pairwise comparison by asking groups of people to compare items of known size, like distances between cities, weight of objects, etc. Although this scale is linear, the priority P generated by the subsequent mathematical procedures follows a geometric progression in agreement with

TABLE 7.3
Example of scales that can be used to compare two needs

	Perception		Stimulus (Priority Ratio)	
Level of perception intensity	Perceived priority ratio	Saaty, 5 Steps	Saaty, with Intermediate Steps	2^n Geometric Progression
4	Extreme	9	9	16
	Very strong to extreme		8	
3	Very strong	7	7	8
	Strong to very strong		6	
2	Strong	5	5	4
	Moderate to strong		4	
1	Moderate	3	3	2
	Equal to moderate		2	
0	Equal	1	1	1

Weber–Fechner law. Alternative scales have been proposed during subsequent years [3]. A simple one is the geometrical progression $S_n = 2^n$ (Figure 1.4). If it is used, the priority ratio between two elements with perception intensity n and m is:

$$\frac{S_n}{S_m} = 2^{n-m}. \qquad (7.2)$$

Saaty's scale and the geometrical progression are compared in Table 7.3. It is remarkable that, despite differences, they provide very similar results (i.e. local priorities) when used in calculations. The reason is that the algorithm sees the 1–9 scale as an approximation of the geometrical progression and generates the same output. This equivalence between scales is further explained with an example in Appendix B.

7.4   CALCULATION

7.4.1   BASIC THEORY

Consider a set of N elements $E_1 ... E_N$ you want to prioritize (for instance, N customer needs), and let $P_1, ..., P_N$ be their normalized priorities:

$$P_1 + P_2 + ... + P_N = 1. \qquad (7.3)$$

Priorities are the unknown parameters you want to calculate. Using the pairwise comparison judgments from your audience, you can build the "judgment matrix":

$$A = \begin{pmatrix} 1 & a_{1,2} & \cdots & a_{1,N} \\ 1/a_{1,2} & 1 & \cdots & a_{2,N} \\ \vdots & \vdots & \ddots & \vdots \\ 1/a_{1,N} & 1/a_{2,N} & \cdots & 1 \end{pmatrix}. \tag{7.4}$$

The elements a_{ij} are the approximation of priority ratios you derive from customer answers, typically converting answers in natural language into numbers. For instance:

- $a_{1,2} = 2$ means that E_1 is twice as important as E_2.
- $a_{3,2} = 1/4$ means that E_3 is four times less important than E_2.
- $a_{1,3} = 1$ means that E_1 is as important as E_3.

Note that:

- All the elements on the diagonal are equal to 1 since every element is as important as itself.
- $(a_{i,j} = 1/a_{j,i})$ since, for instance, the importance of E_1 element is twice as important as E_2, then the importance of E_2 must be one half of the importance of E_1.

Therefore, although A has N^2 elements, you just need to determine the $(N^2-N)/2$ elements in the top half or the bottom half of the matrix.

If the answers of the audience were purely logical, you would have:

$$a_{ij} = P_i/P_j, \tag{7.5}$$

and the following "consistency condition" would be satisfied:

$$a_{i,j} a_{j,k} = a_{i,k}. \tag{7.6}$$

The meaning of this condition can be easily understood with an example: if, for instance, E_1 is twice as important as E_2 ($a_{1,2} = 2$), and E_2 is three times as important as E_3 ($a_{2,3} = 3$), then E_1 must be six times more important than E_3 ($a_{1,3} = a_{1,2} \cdot a_{2,3} = 2 \cdot 3 = 6$).

For a consistent judgment matrix, the set $(P_1,...,P_N)$ can be obtained as the solution of the following linear system (you can simply verify it by performing the vector multiplication on the left-hand side) [1, 2]:

$$\begin{pmatrix} 1 & \cdots & \dfrac{P_1}{P_N} \\ \vdots & \ddots & \vdots \\ \dfrac{P_N}{P_1} & \cdots & 1 \end{pmatrix} \cdot \begin{pmatrix} P_1 \\ \vdots \\ P_N \end{pmatrix} = \lambda \begin{pmatrix} P_1 \\ \vdots \\ P_N \end{pmatrix}. \tag{7.7}$$

The solution $\mathbf{P} = (P_1,...,P_N)$ is thus an eigenvector of the matrix A, and λ is its eigenvalue.

For a square matrix of dimension N, there are N pairs of eigenvectors and eigenvalues. How do you know which eigenvector is the solution you are looking for? In general, both the eigenvalue and the components of the eigenvector are complex numbers. We are however looking for a solution with positive real components (all the numbers $P_1,...,P_N$ must be positive to be meaningful as priority), and we can discard all eigenvectors which are complex, or have one or more negative components.

Moreover, A has two important properties:

1. A contains only positive numbers. Consequently, in view of the Perron–Frobenius theorem, there is only one eigenvector whose components are all positive. The corresponding eigenvalue is a positive real number, and it is larger than the absolute value of any other eigenvalue (it is called the "leading" eigenvalue).
2. A is consistent. For a positive consistent matrix, all eigenvalues but one are zero. The eigenvalue which is not null is $\lambda = N$.

We can thus conclude that for a consistent matrix, the solution we are looking for is the eigenvector with eigenvalue $\lambda = N$.

Consider now a more realistic case in which the judgments, which are made by humans, are not completely consistent: in general $a_{i,j} \neq P_i/P_j$. You could hear, for instance, that E_1 is twice as important as E_2, that E_2 is three times more important than E_3, and then E_1 is "just" 1.5 times more important than E_3. Or, in an extreme case, even that E_3 is more important than E_1. However, A remains positive, and the Perron–Frobenius theorem still holds. If the deviation from a consistent matrix can be considered a perturbation:

$$a_{i,j} = \frac{P_i}{P_j} + \gamma_{i,j}, \tag{7.8}$$

where the values $\gamma_{i,j}$ are small in comparison to P_i/P_j, it can be shown that the leading eigenvalue of A is always larger than N ($\lambda \geq N$), and the equality holds only if the perturbation vanishes.

Thus, it is reasonable to conclude that:

1. The leading eigenvector $(w_1,...,w_N)$ of a decision matrix A is the best approximation to the priority vector.
2. The closest the leading eigenvalue is to N, the more consistent the judgments are.

In view of point (2), the consistency of the decision matrix can be quantified using the difference between λ and N. One can look, for instance, at the "consistency ratio" (CR):

$$CR = \frac{(\lambda - N)}{RI(n) \cdot (N-1)},$$ (7.9)

where RI(n) ("random consistency index") is an empirical value which depends on the dimension of the matrix. RI values for matrices with dimension from 2 to 7 are reported in Table 7.2. If CR is larger than 0.1, judgments are likely to be significantly inconsistent, and one should try to work with the respondent to identify and solve the inconsistency.

7.4.2 NUMERICAL METHODS

Software tools to calculate the leading eigenvectors and eigenvalues of the decision matrix can be easily found on the internet. Many alternatives are available, ranging from dedicated commercial AHP applications, which offer useful tools to organize decision data, to simple add-ins for spreadsheets.

There are algorithms simple enough to be implemented with a spreadsheet without special skills too. They are [4]:

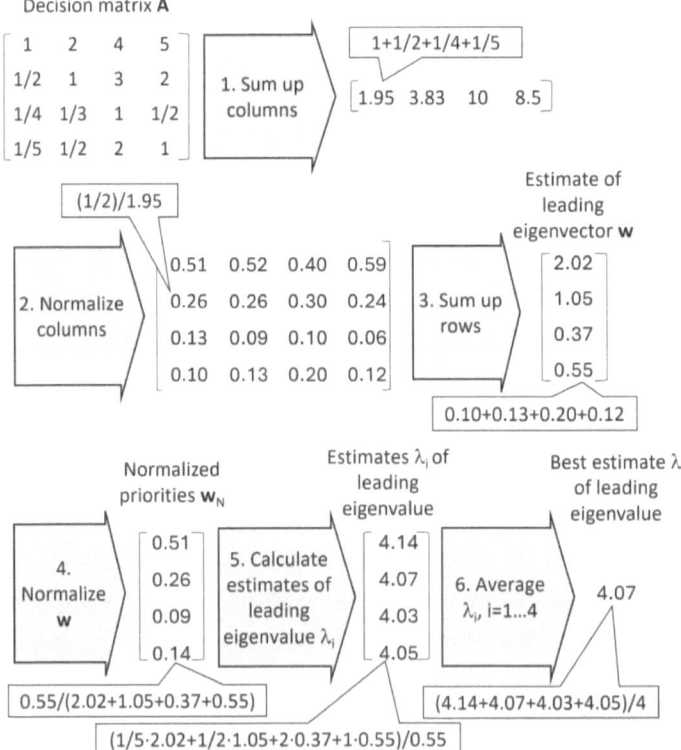

FIGURE 7.1 Example of leading eigenvector calculation using the row average of normalized columns (RANC) method.

1. Calculate the geometric average of the rows of the decision matrix, then normalize the resulting vector to 1. More in detail: the nth component w_n of the leading eigenvector **w** is obtained as the geometric average of the elements in the nth row of the decision matrix **A**. The normalized weights are obtained by dividing each component of the leading eigenvector by the sum of all components.
2. Row average of normalized columns. This method is illustrated in Figure 7.1. The first step is to normalize the column of the decision matrix **A** by dividing each column by the sum of its elements. The second step is to obtain the approximate components of the leading eigenvector **w** by adding up all the elements in the same row. Finally, the normalized priorities $\mathbf{w_N}$ are obtained by normalizing **w**.

With both methods, if you want to calculate the consistency ratio, you need the leading eigenvalue too. Approximations λ of the leading eigenvalue are obtained by multiplying the ith row of the decision matrix by the leading eigenvector, and by dividing the result by the ith component of the eigenvector. In general, different approximations are obtained by taking different rows ($i = 1, 2, 3...$), as shown in Figure 7.1. It thus makes sense to take the arithmetic mean of these values to obtain a best estimate of the leading eigenvalue λ. In the example of the figure our best estimate of the leading eigenvalue is thus $\lambda = 4.07$ which, with $N = 4$, gives a consistency ratio $CR \cong 0.03$.

It can be easily verified with a direct calculation that these two methods generate exactly the leading eigenvector if the matrix is consistent. In all the cases of practical interest, including decision matrix which are somehow inconsistent, the leading eigenvector is anyhow obtained with sufficient accuracy.

7.4.3 AVERAGES

The pairwise comparisons $a_{ij}^{(1)}$, $a_{ij}^{(2)}...a_{ij}^{(M)}$ of M customers can be combined using their geometrical mean $<a_{ij}>$:

$$a_{ij} = \sqrt[M]{a_{i,j}^{(1)} \cdot a_{i,j}^{(2)} \cdots a_{i,j}^{(M)}}. \tag{7.10}$$

You can simply convince yourself with few examples that the geometric mean is the right calculation in this case:

a. Opposite judgments produce a neutral judgment. If, for instance, the first customers says E_1 is twice more important than E_2 ($a_{1,2}^{(1)} = 2$) and the second customer says that E_2 is twice more important than E_1 ($a_{1,2}^{(2)} = 1/2$), the geometric mean produces a neutral result:

$$a_{1,2} = \sqrt[2]{2 \cdot 1/2} = 1. \tag{7.11}$$

b.   Many identical judgments produce an identical mean. Consider for instance 3 customers saying that E_1 is twice more important than E_2:

$$a_{1,2} = \sqrt[3]{2 \cdot 2 \cdot 2} = 2 \cdot \qquad (7.12)$$

c.   If the judgments are consistent, the mean of ratios $a_{1,2}^{(1)}, a_{1,2}^{(2)}...a_{1,2}^{(N)}$ is the ratio of means:

$$a_{1,2} = \sqrt[N]{\left(a_{1,2}^{(1)} \cdot ... \cdot a_{1,2}^{(N)}\right)} = \sqrt[N]{\left(\frac{P_1^{(1)}}{P_2^{(1)}} \cdot ... \cdot \frac{P_1^{(N)}}{P_2^{(N)}}\right)} = \frac{\sqrt[N]{P_1^{(1)} \cdot ... \cdot P_1^{(N)}}}{\sqrt[N]{P_2^{(1)} \cdot ... \cdot P_2^{(N)}}} = \frac{P_1}{P_2}. \qquad (7.13)$$

Condition (b) would be satisfied by the arithmetic mean too, but (a) and (c) would not.

7.5   EXAMPLE 1: APPLES AND ORANGES

Let's assume that you enroll three groups of people, each with the same number of members, to judge the tastiness of different fruits. Members of the first group think in a very logical way; those in the second group think like an average human, and those in the last group are particularly incoherent. Since $N = 3$, you need $3 \cdot (3 - 1)/2 = 3$ questions when you interrogate your population:

- Which is the tastier between apples and oranges? How much?
- Which is the tastier between apples and blueberries? How much?
- Which is the tastier between oranges and blueberries? How much?

You start with the logical guys. First, you perform a pairwise comparison with natural language, then you translate answers into ratios using an empirical conversion scale and finally you average the ratios provided by different group members. The result is as follows:

- Apples are two times tastier than oranges.
- Apples are six times tastier than blueberries.
- Oranges are three times tastier than blueberries.

The judgment matrix is:

$$\begin{pmatrix} 1 & 2 & 6 \\ 1/2 & 1 & 3 \\ 1/6 & 1/3 & 1 \end{pmatrix}.$$

No surprises, this matrix is perfectly consistent (CR=0). The only non-null eigenvalue is 3, and its normalized eigenvector is (0.6, 0.3, 0.1). It means that apples have 60% tastiness, oranges 30%, and blueberry 10%. Of course, when the matrix is consistent, the result appears trivial. However, it is unlikely that you meet in real life a person with perfectly consistent judgment.

The "average" persons could provide a response like the following:

$$\begin{pmatrix} 1 & 2 & 5 \\ 1/2 & 1 & 4 \\ 1/5 & 1/4 & 1 \end{pmatrix}.$$

This group either underestimates the tastiness of apples over blueberries or overestimates the tastiness of apples over oranges. However, there are not severe inconsistencies in their judgment: the leading eigenvalue is 3.03 and CR=0.02. The normalized leading eigenvector is (0.57, 0.33, 0.1), which is very close to what you got from the logical guys.

Finally, when you ask the "incoherent" group, you get:

$$\begin{pmatrix} 1 & 2 & 1/3 \\ 1/2 & 1 & 4 \\ 3 & 1/4 & 1 \end{pmatrix}.$$

These people are telling you that they prefer apples over oranges, and oranges over blueberries, but blueberries over apples, which sounds inconsistent. Now the leading eigenvalue is 4.23 and the consistency ratio is very high (CR=1.06>>0.1). The leading eigenvector is (0.29, 0.41, 0.30). The order of priorities is completely different from other groups.

As a final part of the exercise, let's assume that the "incoherent" interviewees confirm their judgment, and that you decide they belong to the same population of the other two groups and it makes sense to average all results. When you average the responses of all respondents you get:

$$\begin{pmatrix} 1 & 2 & 2.15 \\ 0.5 & 1 & 3.63 \\ 0.46 & 0.28 & 1 \end{pmatrix}.$$

The leading eigenvalue is 3.17, and the leading eigenvector is (0.49, 0.36, 0.15). The average process mitigates the effect of the incoherent person, and the order of priorities is correct. However, the distances between the three priorities are still very compressed, and the consistency ratio (CR=0.14) is higher than what you would usually accept.

7.6   EXAMPLE 2: REGULAR JOE'S CASE STUDY

Consider again the hierarchy diagram of the "regular Joe" case study (Figuer 5.4). It exhibits ten groups of needs, distributed in four levels (from primary to quaternary level).

TABLE 7.4

Example of possible judgments provided by ten customers interrogated during regular Joe's case study

Customer No.	Importance of "I need to be relaxed when using my car" vs "I need to improve my image" (Natural Language)	Customer judgment converted to score
1	Strongly more important	4.00
2	Very strongly more important	8.00
3	Strongly more important	4.00
4	Equally important	1.00
5	Strongly more important	4.00
6	Very strongly more important	8.00
7	Moderately more important	2.00
8	Moderately/strongly more important	3.00
9	Equally important	1.00
10	Moderately less important	0.50
	Geometrical mean	2.56

The prioritization of each group with pairwise comparison requires its own judgment matrix. Calculations are in this case very simple because all groups contain just two needs. Consider, for instance, the group at primary level, which has the needs: "I need to be relaxed when using my car" and "I need to improve my image". Assume that you interrogate ten customers who provide the answers presented in Table 7.4.

If customers judgments, provided using natural language, are converted to a score, we find that on the average "I need to be relaxed..." is 2.56 times more important than "I need to improve my image". This is the only entry needed in the 2×2 judgment matrix:

$$\begin{pmatrix} 1 & 2.56 \\ \dfrac{1}{2.56} & 1 \end{pmatrix}.$$

The normalized components of the leading eigenvector are the priorities of the two needs:

$$\begin{pmatrix} 0.72 \\ 0.28 \end{pmatrix}.$$

We see that the first need is $0.72/0.28 \cong 2.6$ time more important to satisfy than the second need.

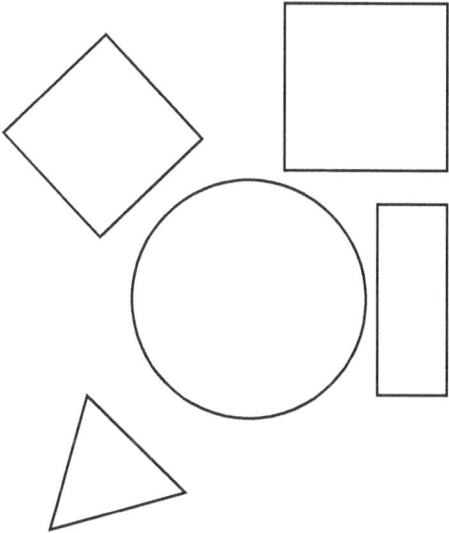

FIGURE 7.2 Example of shapes which could be used to test the accuracy of pairwise comparison method.

7.7   CHECK YOURSELF THE ACCURACY OF AHP

You can check for yourself the accuracy offered by the pairwise comparison. This could support your decision whether to use it or not. All you need is a set of elements to compare with a measurable property, like the areas of geometrical shapes [5]. Consider as an example the shapes in Figure 7.2. The naked eye can infer the correct sequence of increasing areas: triangle, rectangle, diamond, square, and circle. However, it is almost impossible to quantify the relative sizes of shapes with numbers without a ruler.

The capability to provide a quantitative evaluation just based on subjective judgments is the power of pairwise comparison. To perform your own validation, follow these steps:

1. Select a judgment scale from Table 7.3.
2. Draw a set of geometrical shapes and measure their area. In our examples, areas are indicated in Table 7.5.
3. Gather a group of colleagues, and make a group pairwise comparison of these shapes: with five shapes the team will have to answer ten questions, like: "what is the biggest area: the circle or the square? And by how much?"
4. Average (using the geometric mean) the responses and calculate the normalized leading eigenvectors.
5. Optionally, ask your colleague to make a direct estimate of areas by assigning a score to every figure. Average the scores (with arithmetic mean) and normalize them.

TABLE 7.5
Areas of the shapes shown in Figure 7.2

Shape	Area (Normalized)
Circle	0.39
Square	0.24
Diamond	0.18
Rectangle	0.12
Triangle	0.07

6. Compare the eigenvectors components from pairwise comparison (and optionally the result of direct estimate) to the true result.

7.8 PRACTICAL HINTS

All the general guidelines and the recommendations provided for simplified prioritization in the previous section are valid for the pairwise comparison also. Best practice at the time of writing this book was to perform the prioritization during an on-site meeting or a video call with the customer, to minimize the risk of misunderstanding, and give attention to the homogeneity of customer segment.

The authors acknowledge that the evolution of digital communication may create new scenarios in which offline alternatives, without direct communication between the interviewer and the interviewee are necessary, or convenient. For instance, the prioritization SW, due to analytic or artificial intelligence methods, could take over from the interviewer the task of moderating the prioritization. The possible use of artificial intelligence in product design is discussed in more detail in Chapter 13.

Besides that, with pairwise comparison time becomes even more important than usual, and it is essential that you are well organized. First, check the hierarchy diagram before starting interviews. Try to avoid groups with more than four needs. Large groups are not suitable for pairwise comparison since they require ten or more questions.

Carefully select the natural language scale you want to use. If possible, offer verbal judgments in the native customer language.

Special care is needed to identify and solve inconsistencies. Ideally, you should be able to verify the consistency of answers during the prioritization. Thus, be sure that the judgments can be automatically processed when you record them. For instance, prepare in advance the spreadsheet with all necessary comparisons and formulas, or use a dedicated AHP software.

If consistency of answers within a group is not achieved, try to understand with the customer what is inconsistent. Resolve the inconsistency if it is due to the lack of attention, or due to a misunderstanding. There are commonly four reasons for high levels of inconsistency; the first three are the responsibility of the person conducting the interview, not the respondent.

- When entering the judgment $a_{i,j}$, it is erroneously recorded as the reciprocal $1/a_{i,j}$. Dedicated AHP applications make this mistake easier to avoid, but spreadsheet users should take care.
- When the customer is asked to compare need 1 with need 2, they mistakenly just rate the importance to satisfy need 1 rather than comparing how much more important need 1 is to need 2. To avoid this misunderstanding, you can rephrase a few comparisons in the beginning and ask a confirmation, for example, "We understand that for you Need 1 is extremely more important than Need 2. Is it correct?"
- The customer need is imprecisely formulated so that its meaning in the mind of the customer depends on what need it is compared with. In such a case, the team should phrase their customer need statements more crisply.
- The customer is actually inconsistent in his/her judgment. If, despite your attempts to identify and solve the inconsistency, it turns out to be intrinsic to the way of thinking of the customer, accept it.

If you are using a spreadsheet, you will need a matrix for each customer (to verify the consistency of his answers) and a "master decision matrix" for the average scores. Remember to average the scores corresponding to natural language (e.g., 1, 3, 5, 7, 9) using the geometric mean, not the arithmetic mean. Dedicated SW tools usually take care of data management for you.

BIBLIOGRAPHY

[1] Saaty T. L., "A scaling method for priorities in hierarchical structures", *Journal of Mathematical Psychology*, vol. 15, no. 3, pp. 23481, June 1, 1977.
[2] Saaty R. W., "The analytic hierarchy process-what it is and how it is used", *Mathematical Modelling*, vol. 9, no. 3–5 pp. 161–176, 1987.
[3] Goepel K. D., "Comparison of judgment scales of the analytical hierarchy process — a new approach", *International Journal of Information Technology & Decision Making*, vol. 18, pp. 445–463, 2019.
[4] Saaty T. L., *Decision making for leaders: The analytical hierarchy process for decisions in a complex world*, University of Pittsburgh, Pittsburgh (PA), 1988.
[5] Whitaker, R., "Validation examples of the analytic hierarchy process and analytic network process", *Mathematical and Computer Modelling*, vol. 46, no. 7, pp. 840–859, October 1, 2007.

8 Create a Product Concept

8.1 PURPOSE

Due to visits and interviews with customers, at this point you have understood what goes right and what goes wrong in their lives when they use existing products and methods. These data have been annotated in the customer process model, and further analyzed with the customer visit table and the customer voice table. This analysis allowed you to understand which are the unsolved needs of customers as well as important needs that have been already solved by products available on the market. Then, you have structured the unsolved needs and prioritized them together with the customers. It is now time to use this knowledge to draft the product concept.

Here we enter the realm of creativity. Up to now the proposed method has been very analytical, with the aim of understanding the customers. Suggestions about how to evade your paradigms and to enhance creative thinking are given in Chapter 12. The purpose of this chapter is to provide tools to guide the creation of a product concept which solves the most important customer needs.

By "product concept" we mean a high-level product description that you can show to customer to double-check this is what they would buy, to the management to have the development approved, and to the different departments which involve in product lifecycle so that they understand what must be done.

It is key to focus on top customer needs. Being able to address all unsolved needs revealed during market research is illusory and a product design should be focused to solve a few "top" needs, that is, the ones customers prioritize as most important to improve in their work or lives. As a rule of thumb, you can hope to satisfy at most three needs, but it should be clear which is the top one, so that a balance or trade-off can be performed in case of a bottleneck of time, budget, or resources. In some cases, solving this engineering bottlenecks can lead to disruptive and game-changing innovations.

A key factor to select the need to be solved is the customer priority, that is, the output of need prioritization discussed earlier. However, other aspects must be typically considered. These might include the strategic value of an emerging technology for a company, the effect on profitability, the availability of necessary know-how in your company. For instance, it could be very risky to try solving a need if it requires key technologies already mastered by a competitor: the chances to fail are high and,

DOI: 10.1201/9781003544845-8

if it happens, you will provide the competitors with hints they need to quickly develop an even more successful product. Conversely, the key technology may be one that it is imperative to develop to remain a player in the industry, which could lead to forming exclusive partnerships with new suppliers.

8.2 CONTENT OF PRODUCT DESIGN

The detailed content of a product design depends on the nature of the product. The product may be, for instance, a physical object or a service. It may include hardware or software components, or specific information contained in databases. If it is a service, human activity is likely part of the product too. However, in general, product design can be seen as a set of related requirements.

The importance of properly dealing with requirement is well understood in modern industry: requirement engineering is a discipline on its own (see, e.g., [1]), and it is the subject of many standards and guidelines. Your company probably already has experts and one or more powerful software (SW) to manage requirements, and it is not our goal to provide an alternative to your existing procedures. Instead, our purpose is to equip you with simple tools that are helpful to outline a product concept satisfying top customer need. Later, if the product development is approved, your concept can be translated into the requirement management system of your company.

The individual requirements and their interdependencies can be described with text, pictures, and many complex data structures depending on design complexity, the level of abstraction, the nature of the product, etc. In this book we will mostly use text tables for simplest cases, and trees otherwise.

8.2.1 FUNCTIONAL AND NONFUNCTIONAL REQUIREMENTS

Requirements can be broadly classified into two categories, as summarized in Table 8.1:

- Functional requirements describe what the product must do and refer to parts of the product accessible to the user. They are directly related to the voice of customers and to their desires. For these reasons the implementation of functional requirements is typically mandatory. Functional requirements are often described by "use cases" (and in SW development by user stories) expressed in the form "the product must do ..." or "I as the user can...", etc. Depending on the context, these requirements are called features, functions, characteristics, etc. A customer need can be considered a sort of "fundamental" or "master functional requirement" too.
- Nonfunctional requirements on the contrary describe how the product works. They are used to impose constraints on the design by specifying performances, security level, reliability, etc. They relate to the expectations of the customers and are usually negotiable, as alternatives can be proposed by engineers (unless strict norms apply). Nonfunctional requirements are also called "quality attributes", "quality goals", "constraints", "technical requirements", and "enabling technologies".

TABLE 8.1
Comparison of functional and nonfunctional requirements

Requirement characteristic	Functional	Nonfunctional
Requirement content	What the product does (features, functions, characteristics)	How the product works (performances, technical constraints, enabling technologies, system architecture, etc.)
Corresponding item in customer life	Mostly wishes	Mostly expectations
Who write them?	It is a translation done by the product owner/manager of a customer need	Developers
Negotiability?	Very limited	Yes, usually they are one of several alternatives (unless strict norms apply)
Can be tested by the user (at least in principle)	Yes	No

The same requirement can be functional or nonfunctional depending on the context (and this is typically the case of performance): the maximum speed of a car can be a functional requirement for a sport car, and a nonfunctional one for family wagons.

8.2.2 TRACEABILITY OF REQUIREMENTS

Any real design contains many requirements, with a different degree of generality, which form a sort of hierarchy:

- High level requirements: these are directly responsible for satisfying a customer need. They are typically broadly stated functional requirements. If they are nonfunctional, they describe fundamental aspects of the product architecture, basic performances, or constraints with a direct impact on customer experience.
- Low level requirements: these are necessary to implement the high level requirements. They are typically very detailed, nonfunctional requirements. If they are functional, they describe accessory functionalities, or functionalities that can be negotiated and replaced with alternatives.

Creating a design requires a clear understanding of the causal relationships between needs and requirements, and between requirements at different hierarchy levels. You must be aware of "why are we doing something" so that you can take decisions like what can be removed to avoid overengineering; what must be added to fulfill

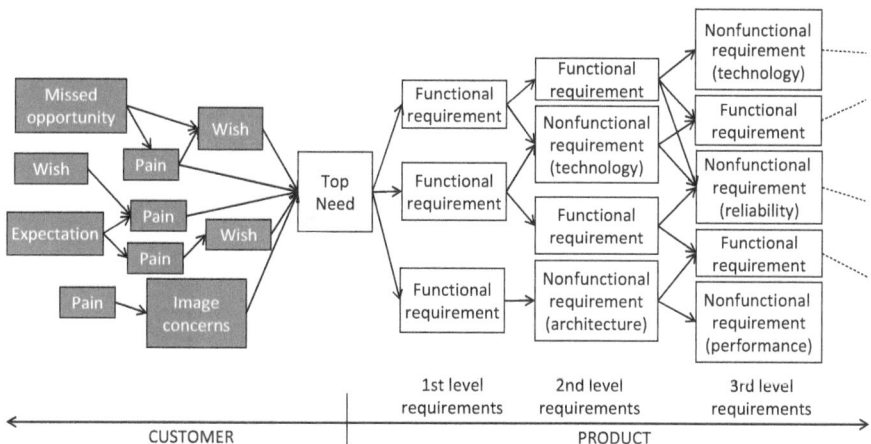

FIGURE 8.1 Sketch showing (i) the role of a need in pain relief and wishes fulfillment and (ii) the way in which it is satisfied by product functions and features.

top needs; what can be replaced by something else to optimize costs; what must be adapted after a change, etc. In other words, you must understand the traceability of requirements to needs. The concept of traceability is shown in Figure 8.1, where the mind of the customer can be seen as a complex network of interconnected thoughts related to pains, wishes, expectations, image concerns, missed opportunities, etc. The "need" rationalizes in a condensed form this complexity. It is a summary of customer life and at the same time the master functional requirement of the product. Requirements at highest level ("1st level requirements") are directly responsible of solving the need. Requirements at the second level are necessary to implement the requirements at the first level, etc. In general, when moving from a high level to a low level, requirements become more detailed, more numerous, and less functional.

8.2.3 An Example

Consider as an example the design of a mobile phone for people making frequent business travels between Europe and North America. A top need of them could be "I need to use the same mobile phone in US and Europe". This need could be solved by two high-level functional requirements: "The product can be used in US" and "The product can be used in Europe", which are the parents of many other requirements, as shown in Figure 8.2. There will be of course many other needs to be solved, for instance, those related to the use of the device (phone calls, video call, internet navigation, video streaming, etc.) which are not shown in this sketch. Other requirements in the hierarchy describe product features related to its usability, for instance, where the phone must be used (big urban areas with wide network coverage, countryside, etc.). All these requirements are hardly negotiable, since if they are not satisfied the product is useless for the customer. They can be easily tested by customers on their own too. To implement these functional requirements, many nonfunctional requirements will

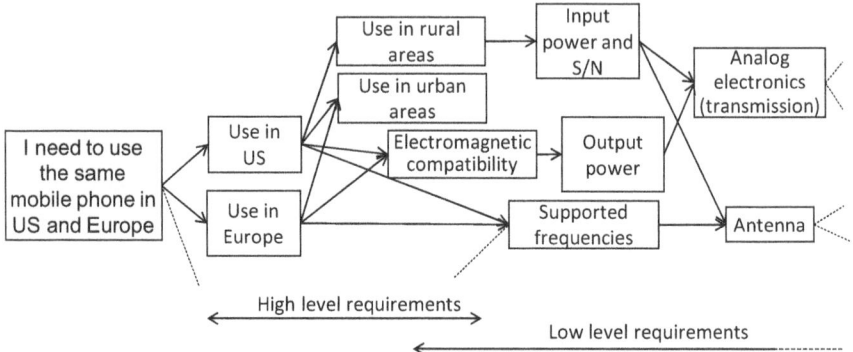

FIGURE 8.2   Example of requirements solving the need "I need to use the same mobile phone in US and Europe".

have to be satisfied, for instance, the set of supported frequencies and the design of the antenna. There will probably be several possible combinations of frequencies and antenna design, each with pros and cons, and the developers will have to identify the best alternative. Customers do not care about this selection: they usually know nothing about the frequency spectrum of mobile communication, and they have not seen an external antenna since 2010 or earlier. They just expect the device to work.

8.3   TABLES

The easiest way to establish a causality link between customer needs and requirements is to use tables. This is possible if the design is relatively simple. In the example presented in Table 8.2 the product must satisfy three needs. Every need is addressed by a function which must satisfy a certain standard and can be implemented with one feature requiring one specific technology. This is clearly an extremely simple case.

A clear but more realistic case study can be derived from regular Joe's car example. We start from the need prioritization of Figure 6.9, and we assume that we want to focus on the top need: "I can easily find where I parked my car". The result of our imaginary product design team is shown in Table 8.3. In particular, the top need could be solved by tracking the parking position with a mobile phone. The implementation of this function requires two main features:

1. An interface between the car's electronic system and the mobile phone, which can be realized via Bluetooth or Wi-Fi.
2. The actual application, which is pure software. It shall run at least on iOS and Android operating systems to cover most of the market.

Finally, the app will require at least two other features:

*   An algorithm to acquire the parking position, which could be based on a Global Positioning System (GPS) signal.

TABLE 8.2
Example of how traceability between design elements is achieved with a table in a simple case

Need	Functional requirement	Performances, standards	Technology	...
Need 1	Function 1	Standard 1	Technology 1	
Need 2	Function 2	Standard 2	Technology 2	
Need 3	Function 3	Standard 3	Technology 3	

Note: The empty column on the right-hand side of the table indicates that additional columns can be added as needed to fully specify the design.

TABLE 8.3
Example of car design satisfying "regular Joe's" top need

Need	Requirements (1st level)	Requirements (2nd level)	Requirement (3rd level)	Requirements (4th level)
I can easily find where I parked my car	Park location trackability	Car–mobile phone interface	-	Bluetooth or Wi-Fi
		"Take me to my car" application	Algorithm to acquire parking position	Supported operating systems: iOS, Android. GPS
			App interface to mobile navigator	Google Maps, Apple Maps, Waze

Note: The requirement at the first level is functional, and the requirements at the fourth level are nonfunctional.

- An interface to the mobile phone navigation system, to avoid reinventing the wheel by developing ex-novo maps and navigation software. If the interface is suitable to communicate with Google Maps, Apple Maps, and Waze, a large market coverage is ensured.

You can use one or more than one table to establish a need-design causality. For instance, the Blitz QFD® method recommends using two tables: one for the unsolved needs ("maximum value table") and one for the needs already solved by your or competing products that must be solved in the new design too ("expected value table").

If one or more tables are sufficient to carry out the design, there is no need to use more complex tools. There are, however, practical circumstances in which the use of tables is cumbersome. Consider, for instance, the case in which one functional requirement satisfies more than one need; or the situation in which the satisfaction of one need requires the implementation of multiple requirements; or the case in which two functions can solve the same need and must be evaluated as alternatives. In these cases, matrices may enable a more flexible and sophisticated analysis.

8.4   QUALITATIVE MATRICES

Decision matrices can help to capture the correlation between many input and output parameters, and to compare alternatives suitable to solve a problem. Matrices can be both qualitative and quantitative.

As an example, let's assume that we want to address the six main needs of regular Joe's case study. These needs are listed in the first column of the matrix shown in Table 8.4, in order of decreasing importance from top to bottom. Seven high-level functional requirements are considered to satisfy these needs:

- "Park location trackability" remains the solution of the top need: "I can easily find where I parked my car".
- "Passengers' ability to play media", including music, movies, and videogames, could solve the need "My kids will not disturb me from the back seat". It could also convince guest passengers that "I have a taste for music."
- "Back seat roominess" could contribute to avoiding disturbance from kids, who will have more room to play on their own. Additionally, it would provide a comfortable, "classy" sitting to adults, who will think that "what I buy never goes out of style". The same results could be obtained with "interior luxuriousness".
- "Music audibility outside vehicle" with an outside loudspeaker would force pedestrians to notice the driver and, with an appropriate selection of tunes, to think that "I have a taste for music".
- "Memorability of car body design" could have a few positive effects: to be noticed by pedestrians as well as by other drivers. Additionally, a proper body design could reinforce the feeling that "what I buy never goes out of style".
- Finally, the degree of acceleration would help to "be noticed by other drivers".

This simple example is sufficient to highlight the main benefits offered by matrices:

- Show alternatives.
- Describe how a single requirement can solve more needs.
- Describe how the solution of a single need may require multiple requirements.

The example shows the major, intrinsic drawback of matrices too: only two design parameters (in this case needs and first level requirements) can be correlated. Every additional parameter will require a new table, with a consequent increase of documentation complexity. For instance, an additional table is necessary to correlate the first requirements with the second level requirements.

TABLE 8.4
Matrix showing the correlation between six top needs and seven 1st level requirements suitable to implement them

Top needs	Park location trackability	Passenger ability to play media	Back seat roominess	Interior luxurious design	Music audibility outside the vehicle	Memorability of car body design	Acceleration
			Requirements (1st level)				
I can easily find where I parked my car	✓						
My kids will not disturb me from the back seat		✓	✓				
I want to be noticed by pedestrians					✓	✓	
People think what I buy never goes out of style		✓		✓		✓	
People think I have a taste for music					✓	✓	
I want to be noticed by other drivers							✓

TABLE 8.5
Matrix showing the correlation between the first level requirements and the second level requirements necessary to implement them

First level requiremnts	Second Level Requirements				
	Car–mobile phone interface	"Take me to my car" app	Screen(s) and console for back seats	Move seats back	Increase wheelbase
Park location trackability	✓	✓			
Passengers' ability to play media	✓		✓		
Backseat roominess				✓	✓

Let's try to define the product second level requirements in regular Joe's example (Table 8.5). For the sake of simplicity, we decide to focus on the first two needs so that we have to implement three 1st level requirements only. Here we see that:

- "Park location trackability" can be implemented with two main features ("Car-mobile phone interface" and "Take me to my car app") as discussed in the previous section.
- Passengers in the back can play media and videogames if screens and a gaming console for back seats are provided. The interface between the car and the mobile phone could be useful in this case to bring favorite media and games into the car.
- "Back seat roominess" can be increased in two ways: moving the seats back (thus reducing trunk volume) or by increasing the wheelbase, making the car longer and preserving the trunk volume.

Here we see an additional drawback of matrices: it is not straightforward to distinguish between mandatory features (both "car-mobile phone interface" and "take me to my car app" must be implemented to allow park position tracking) and alternatives (the room around back seats increases if at least one feature between "move seats back" and "increase wheelbase" is developed). This distinction can be achieved by adding additional symbols to the matrix, which however increases the complexity further.

The procedure we used to link with matrices needs and first level requirements, as well as first and second level requirements, can be iterated through all the levels of requirements needed to satisfactorily define the produce (see, e.g., [2]).

8.5   QUANTITATIVE MATRICES

8.5.1   Weights

You can make the correlation between matrix input (e.g. the needs) and the output (e.g. the first level requirements) of a matrix quantitative by judging how

TABLE 8.6
Example of weights which can be used in quantitative matrices

Perceived Importance (perception)	Weights (Stimulus)		
	Geometric Progression	From 1–9 Scale	ISO 16355
Extreme	1.000	1.000	1.000
Very strong	0.500	0.510	0.518
Strong	0.250	0.252	0.267
Moderate	0.125	0.124	0.135
Weak	0.063	0.065	0.069
None	-	-	0.000

Note: Weights are rounded to the third decimal digits as recommended by the ISO standard for comparison, although two decimal digits are sufficient for the applications discussed in this book.

important output parameters are to realize the input parameters. In most cases this judgment is better done by using natural language and a logarithmic scale (in agreement with the Weber–Fechner law) to convert words into weights. These scales are typically normalized in such a way that the largest weight is 1, but this is relevant only for mnemonic reasons, since numbers will be normalized again at the end of the calculation. Scales that are frequently used are shown in Table 8.6: the geometric progression, the "from 1–9 scale", and the scale recommended by the International Organization for Standardization (ISO) standard [3]. The weights "from 1–9 scale" are the same of those in the table of Appendix B, with a different normalization. The weights recommended by the ISO standard, which have been used by the QFD community for decades, are the results of the same eigenvector problem solved to obtain the 1–9 scale, just calculated with a slightly lower numerical accuracy.

According to the spirit of the Weber–Fechner law, the lower weight of geometric progression and "from 1–9 scale" is the non-null smaller stimulus which generates some response. To this regard, the ISO standard is different since it introduces a grade "None" with weight 0. This weight is a deviation from the Weber–Fechner law, since null cannot be obtained as the power of a real number. However, the availability of this score may provide some relief to people uncomfortable giving a finite score (0.06) to a requirement with no importance. If the use of this additional grade helps the team, and if you assign weights based on consensus, feel free to use it: in the end using 0 instead of a very small number will have no influence on the results of the numerical calculation. However, you cannot use the null weight to average individual weights because null would not work with geometric mean.

In the following we will use the geometric progression, which is the easiest to remember, rounded to two decimal digits (which are sufficient for our applications).

TABLE 8.7
Quantitative matrix showing the correlation between needs and functions

Needs		Requirements (1st level)						
Text	Priority (global)	Park trackability	Passenger ability to play media	Back seat roominess	Interior luxurious design	Music audibility outside the car	Memorability of car body design	Acceleration
I can easily find where I parked my car	0.48	1						
My kids will not disturb me from the back seat	0.24		0.5	0.25				
I want to be noticed by pedestrians	0.12					0.5	0.5	
People think what I buy never goes out of stile	0.06			0.25	1	0.5	0.25	
People think I have a taste for music	0.06		0.25			0.5		
I want to be noticed by other drivers	0.03						0.25	0.5
Weight (raw)		0.48	0.14	0.08	0.06	0.12	0.08	0.02
Weight (norm.)		0.50	0.14	0.08	0.06	0.12	0.09	0.02

Strong — Very strong — Very strong — Extreme — Very strong

$(0.5 \cdot 0.25 \cdot 1 \cdot 0.5 \cdot 0.5)^{1/5}$

$0.5 \cdot 0.24 + 0.25 \cdot 0.06$

8.5.2 EXAMPLE OF HOUSE OF QUALITY

As an example, consider again the relationship between regular Joe's needs and high-level requirements. A quantitative analysis is shown in Table 8.7. Besides every need we reported its normalized weight (i.e. the global priority from pairwise comparison prioritization). A matrix showing a quantitative relationship between customer needs and product functional requirements is often called the "House of Quality" in QFD terminology. Such a matrix is often enriched with additional information on the right-hand side, bottom, and on top (which resembles the roof of a house), which makes the displayed information more comprehensive, but more difficult to read [4].

The degree to which requirements satisfy every need is evaluated and reported in the main matrix body. This evaluation can be performed, for instance, by the design team using verbal judgments (extreme, very strong, etc.) which are then converted into a score using one of the scales from Table 8.6. The team can agree on a common judgment. As an alternative, the judgments of individual team members can be averaged. As shown in Table 8.7, five team members judged the correlation between the requirement "Memorability of car body design" and the need "I want to be noticed by pedestrian" as "Very strong", "Strong", "Extreme", "Very strong", and "Very strong", respectively. The geometrical mean of these judgments is 0.5, which corresponds to a "Very strong" judgment.

In this example the relationship between the functional requirement "passengers' ability to play media" (where media is intended to include music, video, and videogames) and "my kids will not disturb me from the back seat" is very strong, since the functional requirement is likely to dramatically mitigate the pain (probably at the cost of kids' education). The same functional requirement has a strong relationship with the need "people think I have a taste for music". Therefore, the raw (i.e. unnormalized) weight of this functional requirement is:

$$0.5 \times 0.24 + 0.25 \times 0.06 = 0.14.$$

From this analysis we see that it is questionable to leave out the functional requirement "Music audibility outside of vehicle". The reason is that it satisfies three minor needs in a very strong way so that its weight is higher than that of "back seat roominess". Thus, these results would require an additional reflection to decide which functional requirements should be implemented at what level of performance.

Let's assume that the design team's analysis confirms that it makes sense to focus on two needs and the three 1st level requirements solving these needs. The quantitative matrix correlating the first level requirements and the second level requirements could look like the one in Table 8.8. Here the weights of the first level requirements are taken from the previous table. The weight of the second level requirement is calculated in the same way as in the House of Quality. For instance, the feature "car-mobile phone interface" has a very strong correlation

TABLE 8.8
Quantitative matrix showing the relationship between the first level and the second level requirements

Requirements (1st level)			Requirements (2nd level)				
Text	Weight		Car–mobile phone interface	"Take me to my car" app	Screen(s) and console for back seats	Move seats back	Increase wheelbase
	(raw)	(norm)					
Park location trackability	0.50	0.70	0.5	1			
Passenger ability to play media	0.14	0.20	0.5		1		
Back seats roominess	0.08	0.11				0.5	0.5
Priority	(raw)		0.45	0.70	0.20	0.05	0.05
	(norm)		0.21	0.33	0.09	0.03	0.03

0.5 0.70+
0.5 0.20

Customer needs		1st level requirements		
Text	Weight (norm.)	Park location trackability	Passenger ability to play media	Back seats roominess
I can easily find where I parked my car	0.48	1		
My kids will not disturb me from the back seats	0.24		0.5	0.25
I want to be noticed by pedestrians	0.12			
People think what I buy never goes out of style	0.06			0.25
People think I have a taste for music	0.06		0.25	
I want to be noticed by other drivers	0.03			

Weight	(raw)	0.48	0.14	0.08
	(norm.)	0.50	0.14	0.08

2nd level requirements				
Car-mobile phone interface	"Take me to my car" app	Screens & console for back seats	Move seats back	Increase wheelbase
0.5	1			
0.5		1		
			0.5	0.5

0.08·0.5+ 0.14·0.50

Weight	(raw)	0.45	0.70	0.20	0.05	0.05
	(norm)	0.21	0.33	0.09	0.03	0.03

FIGURE 8.3 Cascade of two quantitative matrices. The functional requirement weights from the top matrix (needs vs functional requirements) are used in the bottom matrix (functional requirements vs features) for calculations.

with two functions: "park location trackability" and "passengers' ability to play media". Thus, its raw weight is:

$$0.5 \times 0.70 + 0.5 \times 0.20 = 0.45.$$

Linked matrices like these can be shown as a cascade, as shown in Figure 8.3. The cascade can be repeated for several layers of matrices, leading to the complex diagrams typical of comprehensive QFD from the 1970s.

8.6 DESIGN SELECTION WITH MATRICES

Matrices usually offer few design alternatives and one of them must be selected. For instance, in regular Joe's example we identified five 2nd level requirements which could be meaningfully implemented. Shall we implement all these five features, or just a subset? This decision usually requires additional data which is not present in the matrix and must be included in the analysis. Let's develop further the example of Table 8.8. The team decided to consider three fundamental parameters to select the design: "time to market" (which can include development costs), "manufacturing cost", and "is it a selling point?" (i.e. can the feature be directly used to promote the product to customers?). When these additional parameters are added to Table 8.8, Table 8.9 is generated. The team concluded that:

- The interface between the car and the mobile phone requires a reworking of the car's electronic system, which will cost the R&D time and money. However, without this feature the top need cannot be solved, and the cost must be accepted. The interface goes together with the app, which is something nice to show to the customers. The development of the application is completely separated from the development of the car; thus, it will not influence the time to market.
- The screens and the console will significantly increase manufacturing cost; however, they may create excitement if the buyer visits the car dealer with the family.

TABLE 8.9
The first level versus the second level requirements matrix, with additional data about development and marketing

1st level requirements			2nd level requirements				
Text	Weight		Car–mobile phone interface	"Take me to my car" app	Screen(s) and console for back seats	Move seats back	Increase wheelbase
	(raw)	(norm.)					
Park position trackability	0.50	0.70	0.5	1			
Passenger ability to play media	0.14	0.20	0.5		1		
Back seats roominess	0.08	0.11				0.5	0.5
Weight	(raw)		0.45	0.70	0.20	0.05	0.05
	(norm)		0.21	0.33	0.09	0.03	0.03
				0.5·0.70+ 0.5·0.20			
Technical evaluation							
Time to market			☹				☹
Manufacturing cost					☹		☹
Selling point				☺	☺☺	☹	
Include in design?			✓	✓	✓		

Note: The final decision about features to be implemented is captured in the bottom row.

• Moving the seats back is probably not a big challenge; there is however the risk that, since the decision-maker is typically the driver, he will dislike the reduction of available trunk volume. Increasing the wheelbase would mitigate this drawback, at the price of significant development efforts and manufacturing cost.

Now the team could be tempted to include in the design the three 2nd level requirements with highest weight, and to discard the other two, as shown in the bottom row of Table 8.9.

8.7   COMPARISON OF DESIGN TOOLS

In this section we have presented three tools to carry out a product design based on customer needs: tables, qualitative matrices, and quantitative matrices. Every tool has its advantages and drawbacks, which are summarized in Table 8.10.

The main benefit of tables is their simplicity and the fact that they can offer a general overview of the "big picture" because any step of product lifecycle can be represented by a table column (this subject is discussed further in Section 10.1.4).

Qualitative matrices offer a more powerful way to compare alternatives and a fairer comparison between candidates (for instance, all first level requirements are correlated to all needs). On the other hand, the big picture risks are lost. This goes together with the risk of error propagation since important design elements can be forgotten when moving from one matrix to the next.

Quantitative matrices basically exacerbate the pros and cons of qualitative tables: a numerical final output is provided together with a very standardized procedure, at the price of complexity, losing the big picture and being exposed to the risk of error propagation. In particular, the quality of output data is completely dependent on the quality of input data, and common sense corrections are more difficult.

We advise you to choose the simplest method suitable for your project. You should resort to quantitative analysis only if it is necessary, and if trusted, low uncertainty input data are available.

8.8   COMPETITIVE ANALYSIS

The fact that a product satisfies important needs of customers does not automatically mean that customers will buy it. Their response will depend on the product price, which in turn will be compared with what is offered by the competitors. The importance of price cannot be underestimated because it is the only certainty the customer has, the quality being always to some extent hypothetical.

Product pricing is a complex subject, and its detailed discussion is beyond the scope of this book. However, it is worth mentioning a few basic principles that are very important for product design and must but understood. They are sketched in Figure 8.4:

TABLE 8.10
Advantages and drawbacks of the three design tools presented in this section

Method	Complexity	Quality					
		Big picture/error propagation	Consider alternatives	Tell mandatory from optional	Propagate errors	Quantitative output	Fair comparison
Tables	☺	☺	☹	☹	☺	☹	☹
Matrix, qualitative	☹	☹	☺	☹	☹	☹	☺
Matrix, quantitative	☹☹	☹☹	☺	☹	☹☹	☺	☺☺

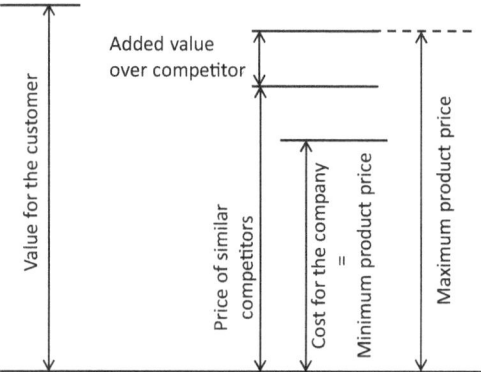

FIGURE 8.4 Sketch showing the relationship between the main variables which determine the range in which product price can be placed.

a. Price cannot be higher than the value perceived by the customer. Value may be determined by looking at what is their problem costing to customers now or what potential gain they could realize with this new product.

b. Price cannot be lower than the cost of the product for the company (considering manufacturing and, if necessary, development, indirect costs, etc.). There may be a short-term strategy to sell at below cost, but this is not sustainable in the long term and, depending on the sales country, it can conflict with competition law.

c. If the product is more expensive than the one by competitors, this additional price must be motivated by an added value. If your product generates more value than the competitor's product because it better satisfies high-priority customer needs, you can consider offering it at a higher price. Remember that the added value must be a real benefit (if competitors are already good enough, being better has no relevance), and that, in view of the Weber–Fecher law, it must be large enough.

A simple example of competitive comparison is shown in Table 8.11. This kind of table is called "quality planning table". Let's assume that the features derived by regular Joe's top needs are added to the existing car "Model A", which will be equipped with a car–mobile phone interface (which allows the use of the "take me to my car" app) and screens and console for passengers sitting in the back. In the table this product is compared to the closest competitors: the car models "Bear" and "Eagle".

In this example the comparison between product concepts is based on the degree to which the product satisfied the two main customer needs and its price. Additional parameters can be added if necessary. The estimated value of solving these needs (i.e., the cost of a problem, or the added value of the solution) for the customer is shown below the table.

TABLE 8.11

Comparison of the selected product design with the closest products offered by competitors

Product	Degree of satisfaction		Price
	I can easily find where I parked my car	*My kids will not disturb me from the back seat*	
Our Model A+ Mobile Interface+ "Take me to my car" app+ Screens and console	☺☺	☺☺	$25,000
Competitor 1 Model "Bear" Full featured		☺	$24,000
Competitor 2 Model "Eagle" Version a56-2			$22,000
Cost of the problem or added value of solution	$1,000	$2,000	

What can be seen from this table is that our product would efficiently solve the two needs. At the same time, it would be the most expensive in its class:

- The "Bear" would partially solve one of the needs, leaving the other need completely unsolved. Thus, the extra price we would ask is likely to be justified.
- The "Eagle" is the cheapest solution available. It does not address the needs at all, but it offers a saving larger than our added value. It would probably be our fiercer competitor.

Note that the design activity outlined in this section is carried out internally by the design team. The table showing the competitive analysis reflects market understanding of internal stakeholders. Before pushing the company for a final decision, it is necessary to talk again to the customers and verify that our design is really creating added value for them, and that the price appears justified. This verification (the "proof of concept") will be the subject of Chapter 9.

BIBLIOGRAPHY

[1] Glinz, M., Loenhoud van, H., Staal, S., and Bühne S. Certified professional for requirements engineering- foundation level- handbook, IREB, 2022.

[2] Mazur, G. H., "QFD in the food processing industry," white paper, QFD Institute, 2008.

[3]   "ISO 16355-1. Application of Statistical and Related Methods to New Technology and Product Development Process- Part 1: General Principles and Perspectives of Quality Function Deployment (QFD)", ISO Standard No. 16355-1, Geneva, 2015.

[4]   Hauser, J. R., Clausing D., "The house of quality", Harvard Business Review, May–June 1988.

9 Proof of Concept

Now that the first draft of product design is ready, you can check with customers and internal stakeholders (typically the level of management above you) if the product you have conceived would be really appealing for them.

At first, you should present the product concept to customers, collect their feedback, and make the necessary adjustments to the design. In this stage you want to understand (1) if the design is really a solution to their needs and (2) if they would pay the price necessary to acquire it. Then, you will present your results internally, explaining the business value of the product, to have the product development approved by management.

9.1 CONCEPT PRESENTATION TO CUSTOMERS

The product concept should be presented to a significant number of customers. A good starting point is to propose a concept presentation to all those who helped you during the collection and prioritization of needs, beginning with the key customer segments and applications. Additional customers can be added if necessary.

You can of course simply present customers with a description of the product you have designed, and then see if customers recognize it as a solution to their pains and needs. It may be worthwhile, however, to provide some context about why you are developing the product and the benefit you want to deliver to customers. One reason for this behavior is that being your concept a high-level description of the product, the customer may be unable to clearly understand the consequences of using the product in their life. Another reason is that this is a good occasion to test possible marketing approaches to be used later at the time of product launch.

The ideal shape of your message to customers depends on the way the marketing and sales teams of your company work. Two examples based on popular marketing and sales approaches are shown in the following. Remember that whatever approach you use, a switch in the organization of your data is necessary to successfully present your concept. Up to this point, data were structured as a sequence of customer problems, customer needs, and product requirements. This structure is optimized to internally explain to engineers why and how they should develop a product. When talking to customers, you should optimize your language and data organization according to their way of thinking.

DOI: 10.1201/9781003544845-9

9.1.1 Presentation Based on Pain–Claim–Gain Approach

A common marketing strategy, often called pain-claim-gain [1], develops in three steps:

a. Make the customer fully aware of their problems, or the opportunities they are currently missing ("the pain"). Of course, a customer always knows in some way their problems, but this knowledge is not always explicit or conscious. You should try to bring the problems you have identified to the surface of customer awareness so that they can fully appreciate your solution.
b. State a "claim" about the product's uniqueness so that it is clear why customers should buy your product instead of a solution from someone else.
c. Explain the customer about how the product would create value for them by solving their problems and enabling their missed opportunities. In other words, which will be "the gain" by purchasing the product. This part typically includes a description of product features.

All the necessary data are already available in the analysis you carried out in previous steps: customer pains were your starting point to understand customer needs, whose solution is the gain for the customer. The claim should be visible from your competitive analysis. What you need to do is to cast this information in an appropriate form.

You can explicitly state the pain, the gain, and the claim; or you can subtly suggest them; or you can let the customers infer them from your presentation. Depending on the customers and on the product, pains and gains could automatically resonate in their mind when you explain the product. In this case there is no need to waste time and bore the customer by explaining what they already know. If the pain is not explicitly mentioned, it could be worth checking indirectly with a few words that the customer is really aware of it. There are unfortunately no fixed rules about how much should be explicitly explained: you must judge case by case, and the answer relies on your understanding of customer psychology and your creativity. We can however provide some helpful advice:

*   Focus on the most important facts and avoid flooding the customer with secondary information. The listener has the tendency to average the information they receive, thus "less is more" for quality products. Presenting unessential data can be useful only in very special cases, for example, if it is necessary to build credibility in front of the customer.
*   Use pictures to capture the attention of the customers and let them absorb information with minimal effort: a picture is literally worth a thousand words. Restrain yourself, however, from putting pictures everywhere and keep the focus on what is important to the customer. You should show one carefully selected memorable picture which the customer can remember, and a few others to make your point.
*   Customers, like other people, are self-important, thus use wording with "you".
*   Provide tangible input. All your statements about the solution and the customer gain should be easy to understand. Be as quantitative as necessary.

- Try to create contrasts (so that the customer can appreciate how their life or work will improve with your product) and to generate emotions (the purchase of the product should be associated with a positive one, of course). If you cannot find a way to generate contrast and emotions with your product design, something important is probably missing in it.
- Be aware that the attention of the customer is maximum at the beginning and at the end of your presentation. If you are showing slides, make good use of the first and the last one. Do not waste the first slide with details about your profile and the fact that your company is "the leader": all your competitors claim to be the leader and all of them hired skilled employers. Customers dedicate their time to you to see how their life or work could improve, not to learn about your new corporate building or your last vacation. Both the opening and the closing should reinforce your claim.

9.1.2   EXAMPLE OF PAIN–CLAIM–GAIN PROOF OF CONCEPT

Let's create a presentation based on regular Joe's car example. We decide to keep the presentation compact by not explicitly mentioning the problems, which will be evoked by the discussion of claims and gains. What we need to show is (see Figure 9.1):

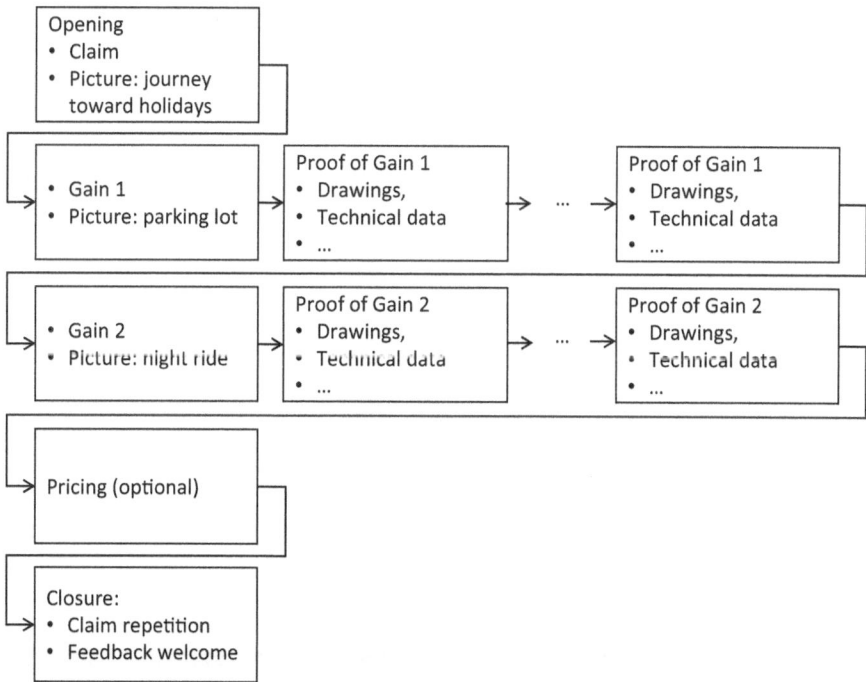

FIGURE 9.1   Possible structure of a proof-of-concept presentation based on pain–claim–gain method.

- The Claim: the car is optimized to provide unprecedented comfort of use and still be affordable for a middle-class family. Of course, one can have a similar or better experience with high-end cars, but not in this price range. That is what makes it unique. A possible claim could be "Comfort was never so affordable". If you are concerned about reminding the customers of their limited economic resources, you may consider "Comfort doesn't have to be expensive" as an alternative to generate softer feelings, at the price of losing the contrast of "never". A picture with a family exiting the car in a sunny sea town with wrinkle-free, tidy clothes and a fresh look could go together with the claim. If the customer is used to exhausting journeys with a cheap car, this image should generate a contrast with their personal experience, and a positive feeling due to the potential solution to the problem.
- Gain 1: a first gain is that the customer will not waste their precious time by looking for their parked car. A catchy image could be a family making their way in a gigantic parking lot under the sun (better avoid showing pounding rain, since it will make cumbersome to use a mobile phone anyway). A couple in the background who are fighting after having lost their way would create a contrast and reinforce the message. In certain cases, it is better to try quantifying the gain. This is easy when the gain is very technical, like reduced fuel consumption or insurance price. In this case you could estimate that the average American family wastes up to 20 hours of free time every year because of parking problems, which are completely solved by your product. It is of course necessary to prove that the gain is provided. At this stage, it can be achieved by showing a sketch of the car–mobile phone connection, and a mock-up (a screenshot) of the mobile phone application.
- Gain 2: children will remain quiet in the back seat even during very long trips. You could show a picture of a family traveling at nighttime with two kids. One is sleeping exhausted, while the other is focused on their screen, maybe with a headset. The driver is fully relaxed, smiles and enjoys music (which may be very important for your customers). Again, a drawing of the entertainment system with minimal technical specification will let the customers clearly understand what you want to provide.

9.1.3   PRESENTATION BASED ON "CHALLENGER" APPROACH

Another popular sales method, especially in the field of business-to-business complex sales, is the so-called "challenger" [2]. It is based on the observation that a fundamental skill of many successful sales representatives is the capability to provide customers "insights". Insights can include unique and valuable perspectives on the market, help to navigate alternatives, help to avoid potential land mines, educate on new issues, etc. This interest in insights reflects a fundamental need of the customer not to buy something, but to learn something.

The goal of a product presentation based on this approach is to make the customer aware of the challenges they are facing, and to let them realize that your product could solve them [2]. Such a presentation can be created by following few steps:

- Build credibility: empathize and show understanding of customers' situation. Do not forget that, even if the insights you provide are largely neutral, in the end your teaching must lead to the unique strengths of your product.
- Challenge customer beliefs and introduce a new perspective. The reaction you are looking for is not "Yes! I agree", but "Huh, I never thought of it that way before", indicating that you made the customer aware of needs he was not yet fully aware of.
- Rationalize with evidence and data the hidden costs of the problem or the missed opportunities. This is the right time to show the data you have collected through your market investigation.
- Create emotions so that the customers see themselves in the story you are telling. Do not let them think: "I am sure it makes a lot of sense for a lot of your customers, but we are different from them."
- Present a new way in which customers could improve their situation, which leads to your product.

9.1.4 EXAMPLE OF "CHALLENGER"-BASED PROOF OF CONCEPT

Regular Joe's car example is not a complex business-to-business sale, thus it would be questionable to apply this method to it in reality. However, we will outline anyhow a challenger proof of concept to provide a clear example. This is the reasoning you could use to challenge the customer and bring it to your new car (Figure 9.2):

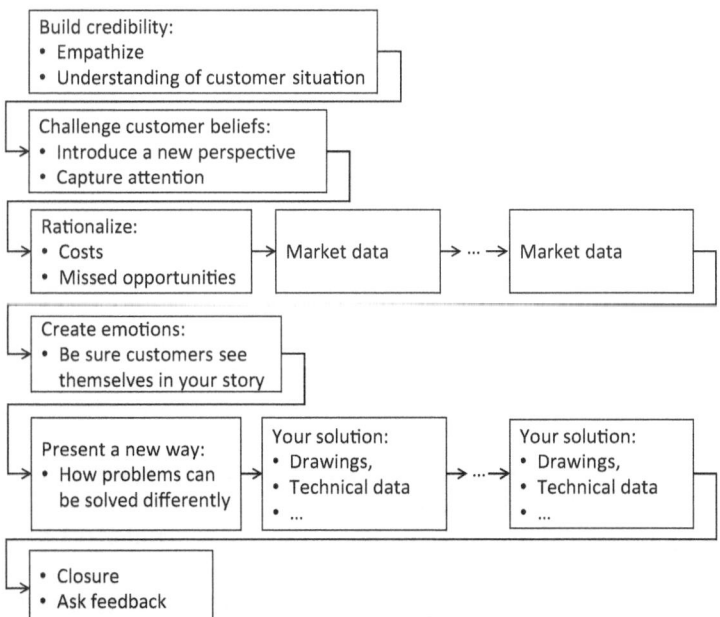

FIGURE 9.2 Possible customer presentation of regular Joe's car concept based on "Challenger" method.

- Empathize: "We know that many fathers need more trunk space to bring home their weekly shopping".
- Challenge: "However, the truck size is not the only factor limiting the comfort of using your car. We know from our market research that there are many other issues. Imagine that it rains, and you cannot find the car in the parking lot or think about kids screaming in the backseats during the way to holidays. Comfort is not a synonym of luxury. "
- Rationalize: "Our investigation shows us that the average American family spends about 20 hours per year searching their lost car in parking lots. There is surely a better way to make use of this amount of time."
- Create emotions: "It could be even worse…it could start raining while you look for your car."
- A new way: "When you look for a new car do not just focus on truck size, but check the overall comfort it can offer. Unfortunately, comfort is often added as an accessory to fast, powerful cars or heavy limousines, and it comes with an additional luxury twist. The reason is that comfort is used by car manufacturers to adjust the price at the highest level a premium customer is willing to pay. However, it would be possible to have affordable and comfortable cars, and it would be the case if they were designed with comfort as their main goal. We offer the "model A" since few years: a reliable car which offers everything you need but the extra comfort. The large number of units sold during recent years already paid out its development, and many improvements to production process allowed us to save the money, which were reinvested in "comfort". This is how the new car will look like…"

9.2 COLLECT CUSTOMERS' FEEDBACK

9.2.1 OVERVIEW

After your presentation, whatever the style you have chosen, you must ask for feedback from the customer to clearly understand how good your idea is. Focus on data you need to resolve your doubts and make decisions: if you already know that a feature will be implemented or discarded, there is no need to discuss it further.

The feedback should be requested right after the product presentation; do not give in to the temptation to later send the customer a written questionnaire. First, since the customer has not seen the real product yet, the risk of later misunderstanding would be very high. Second, you would miss the opportunity to get an in-depth understanding: every time an answer does not fit the big picture you have in mind, you must ask for an explanation. For instance, if a regular Joe-like customer with kids shows no interest for the entertainment consoles in the back, you must ask why: maybe they misunderstood the feature and did not realize the potential of keeping kids busy, or maybe they think that their kids would never spend time in front of a video (and thus they would never buy your product). If the customer does not have time right after the presentation, ask to make a new appointment.

You basically have three groups of questions:

1. At the beginning, questions about the customer situation. This knowledge will help you to contextualize the subsequent answers. Let's look again at regular Joe's example: your customer segments will probably contain mostly male middle-class drivers. To better understand the interviewee, you may want to ask if they are married, the number of children they have, their current car, if they prefer to purchase new or second-hand cars, the purposes for which the car is used, the average distance driven in one year, etc. If you have already talked to the customer during one of previous stages, answers to several questions may be known. In this case it is sufficient to quickly review them with the customer.

2. The middle group of questions will contain standardized, specific questions about the product. They typically include an evaluation of product features, for which you may use the Kano method, outlined in the following section. Examples of additional items you may want to investigate are how the product design solves customer needs, or how it is perceived by the customer in comparison to the offer from competitors.

3. Finally, you can collect any unstructured feedback the customer may want to provide and answer their questions.

9.2.2 How to Formulate Questions

In the end you want to compare the feedback about the product provided by many customers. Therefore, it is important that the questions are the same for all customers, that the customer clearly understands what you are asking, and that their answers let you understand what they have in mind.

The best way to achieve these goals is to use quantitative questions, which are per se the less ambiguous. Examples of quantitative questions are as follows: "what is the typical distance you drive in one year, expressed in km or miles?", "what is your average fuel cost/expenditure per month?", and "what is the average number of years you keep a car?".

If a question cannot be quantitative, try at least to formulate it as a closed-end question with a limited number of predefined answers based on natural language. Carefully avoid ordinal scales like "how many stars out of 5" or "how many points out of 10". Examples of well and badly formulated closed-ended questions are given in Table 9.1. The problem with ordinal scales is that, although you may reasonably assume that what exceeds the middle value is positive, and what is below is negative, you are never sure what is the meaning for the customer of "4 points", or "3 stars", and that this meaning is the same for all customers. Further, attempts to average ordinal rating scores are mathematically meaningless. Why do online shopping apps use them then? The answer is that they work in a different setting, with a different goal. The main benefit of "stars" and "points" is that the scoring is very fast for the customer. Thus, it makes sense to use these scales if you are dealing with many customers who are scoring on their own, and your goal is to maximize the number of respondents. This is the strategy followed by large online stores which want to appear

TABLE 9.1
Examples of well-formulated and poorly formulated questions

Well formulated	Question: How do you judge the statement: "When driving this car my kids will not disturb me from the back seat"?
	Answers: I completely disagree; I somewhat disagree; I somewhat agree; I completely agree.
	Question: Overall, do you think our car will help you remain relaxed while driving?
	Answers: No, it will not help; it will help but not much; it's ok; it would do the job; it exceeds my expectations.
Poorly formulated	Question: On a scale from 1 (complete disagreement) to 10 (complete agreement), how would you rate the correctness of this statement: when driving this car, my kids will not disturb me from the back seat.
	Question: How many stars out of five would you rate our design?

trustable by offering sound scores of products based on the feedback of hundreds or thousands of customers. Our situation, however, is very different: we are probably dealing with 10–30 interviewees, and we want to squeeze out as much information as possible from a one-to-one conversation with them. Thus, there is not any benefit to substituting ratings for natural language.

9.3   INTRODUCTION TO THE KANO METHOD

9.3.1   Customer Satisfaction versus Product Performance

The Kano method [3], [4] is aimed at evaluating how satisfying the customer perceives the presence of a product function, or a certain level of performance. Function or performance should be objective. It can be the weight of a mobile phone, the textile quality of a dress, the presence of a famous actor in a movie, etc. Customer perception on the other hand is subjective by definition and must be asked to the customer.

You may be tempted to ask a customer to evaluate his satisfaction using a single parameter. For instance, you may ask how satisfactory the presence of a certain function (or a certain performance level) is and collect meaningful and useful data in this way. However, the understanding of customer perception will remain superficial because the real relationship between the presence of the function (or performance) and customer satisfaction is typically too complex to be simplified in this way. Let's assume that you have designed a soft drink with a sugar content of 7 g/100 ml. You can let customers taste it and ask them to judge how sweet it is. If they answer "too bitter" or "too sweet", you could know how to improve the product. But what if they answer it tastes good? Do you have margins to decrease the amount of sugar, brand the product as "healthy", and reduce ingredients cost? Or should you make it a little sweeter to match the offer of competitors? To be sure you get useful questions, you

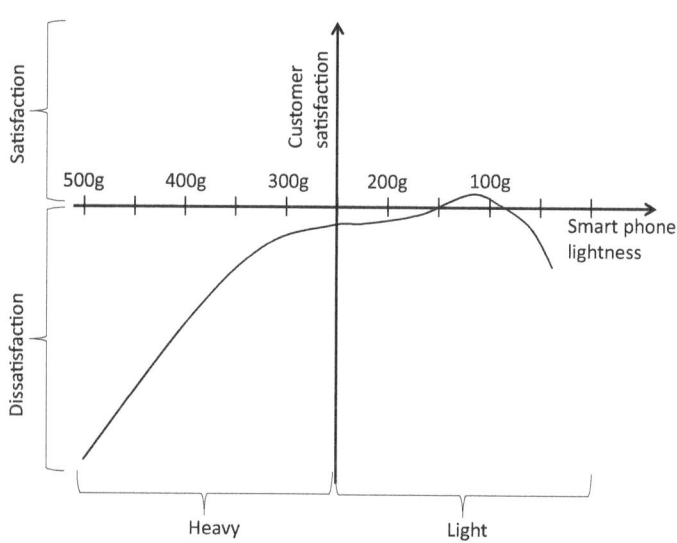

FIGURE 9.3 Hypothetical relationship between weight of a smartphone and customer satisfaction according to the 2023 mobile phone market.

should let the customers taste a few alternatives with different sugar concentrations and see how they react.

As an additional example, consider as a "thought experiment" the feature "smartphone lightness", which can be easily quantified by the weight of a smartphone. If you make an investigation today, you will probably find something like the plot shown in Figure 9.3. In 2023, most smartphones weighted 150–250 g[1], and they were light enough to be carried in a pocket or in a handbag without discomfort. In the past they were lighter than today (in 2007 the first iPhone weighted 135 g, and almost no competitor was heavier than 150 g), but customers have been happy to trade weight for a larger screen and computational power. A 500 g smartphone would surely create huge dissatisfaction. However, a 50 g phone would probably not generate additional satisfaction, simply because we do not care about an extra 50–100 g more in our pocket. On the contrary, a too light phone may cause dissatisfaction because customers cannot recognize whether they have it with them or forgot it somewhere.

9.3.2 MAIN KANO CATEGORIES

The basic simplification introduced by Kano's method, which makes this analysis manageable, is to classify the effect that the presence of a function (or the level of performance) has on customer satisfaction using few typical idealized categories. Three idealized categories are suitable to describe, at least approximately, many real cases (Figure 9.4 and Table 9.2):

1. "Expected feature" (also called "must-be") is what the customer expects as given. When the function exists or the performance level is sufficient, the

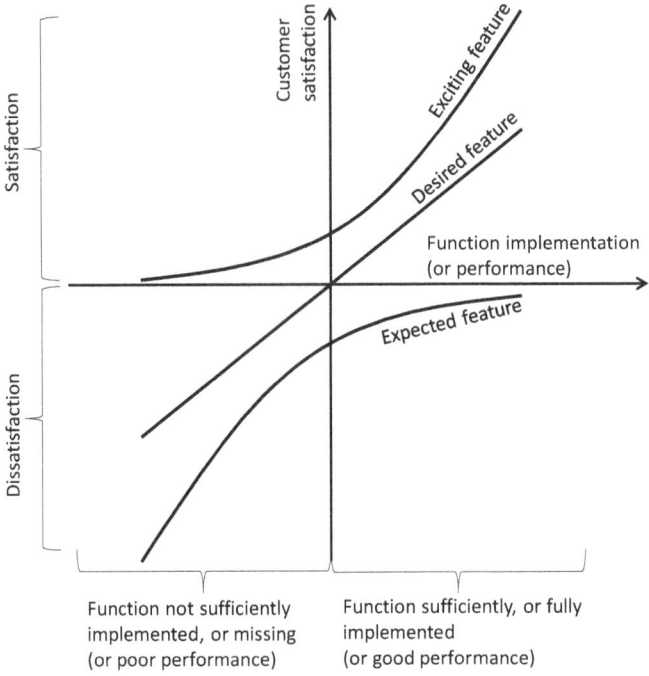

FIGURE 9.4 Relationship between customer perception and performances (or degree of function implementation) in the Kano model.

TABLE 9.2
Feature categories in the Kano model

Feature Category	Spoken?	To be Implemented?
Exciting	No, customers are usually not aware of these features. They get excited only when they see the product (from you or your competitor).	To be determined (e.g. based on cost–benefit analysis for the customer and company)
Desired	Yes. Customers tend to talk a lot about desired features.	Usually implemented
Expected	No. Customers talk about their fundamental expectations only to complain if they see that the product does not fulfil them (usually after a failure).	Must be implemented.

customer usually does not even think of or speak about them. Often, additional satisfaction is not generated if the feature is very good. However, if the function is missing or the performance is insufficient, the customer will be greatly dissatisfied. The lightness of smartphones can be easily considered an expected feature. A missing expected feature is usually recognized after a failure, which for a new product would be too late. Thus, expected features must be identified during design and properly implemented to prevent deep customer dissatisfaction.

2. "Desired features" are what customers usually talk about. For these features the level of satisfaction increases over the performance range. They are also called "one-dimensional" features because they are represented by a straight line in the Kano diagram. For example, consider fuel economy, which is important to most car drivers. Desired features must be usually implemented, but the performance they provide should be carefully tested. Referring to our last example, "fuel economy" requires several trade-offs with weight, engine power, and price, and thus the optimal level of fuel economy to be implemented depends on the target customers. A business car for traveling salesmen may charge a higher price for "very high fuel economy", which would be overkilling for family cars.

3. "Exciting features" generate a large satisfaction if provided; if not, the customer does not care. Customers typically do not talk about these features and realize they can exist only when they see them. For new product developers, this may be too late. An example in 2024 could be a flexible OLED screen for a mobile device. Exciting features can be implemented or not depending on cost/benefit considerations: you must determine if it is worth paying additional development costs and increase product price, to create a strong differentiator against competition. Besides economical consideration, an additional question "shall we be the pioneer" often arises. Try to implement an exciting feature if you are confident that you can market them. Otherwise, be careful since competitors can learn from your failures. One way to double-check whether an exciting feature will sell is to re-examine how important is the customer need that it will solve. A high-priority customer need paired with an exciting feature creates a value proposition to drive sales.

When using Kano categories, it is important to keep in mind few fundamental notions:

a. The category of a feature depends on the customer segment. For instance, "low fuel consumption" is desired by most drivers, but could be exciting for users of heavy-duty trucks, who are used to paying a high price for their gas-guzzling heavy vehicles with powerful engines. It is the customer response to the survey that determines the Kano category, not the organization's perspective.

b. The category of a feature is likely to change with time. For instance, an exciting feature is typically soon copied by competitors and becomes desired. When every competitor offers it, it becomes expected.

c. Kano categories are idealized representations of reality, which is simplified to make its analysis more manageable. For instance, the satisfaction generated by many desired features does not grow indefinitely with performance but saturates at some point. Consider, for instance, the measurement of spatial distribution of dose delivered by radiation therapy: the finest details of such distribution have a typical size of −1 mm, and the spatial resolution of most common two-dimensional (2D) measurement devices is suboptimal to this purpose (3–6 mm), with a few exceptions offering sub-millimeter resolution. Thus, the spatial resolution of a measurement device is a desired feature: the finest, the better. However, a resolution better than 0.1 mm would be overkilling, and it would not generate any additional satisfaction.

9.3.3 OTHER KANO CATEGORIES

The three categories of features presented above are the most common ones. It is however worth to mention few others that can be encountered in practice (Figure 9.5):

- A "skeptical" feature is found when the customers respond they are satisfied if it exists and satisfied if it does not exist. Or if they are dissatisfied in both cases. These responses are usually the results of a poorly worded survey and must be discarded.
- An "indifferent" feature is represented in the Kano diagram by a horizontal line crossing the origin where the two axes meet: it means that the customer does not care about the presence of this feature or its performance. If a feature turns

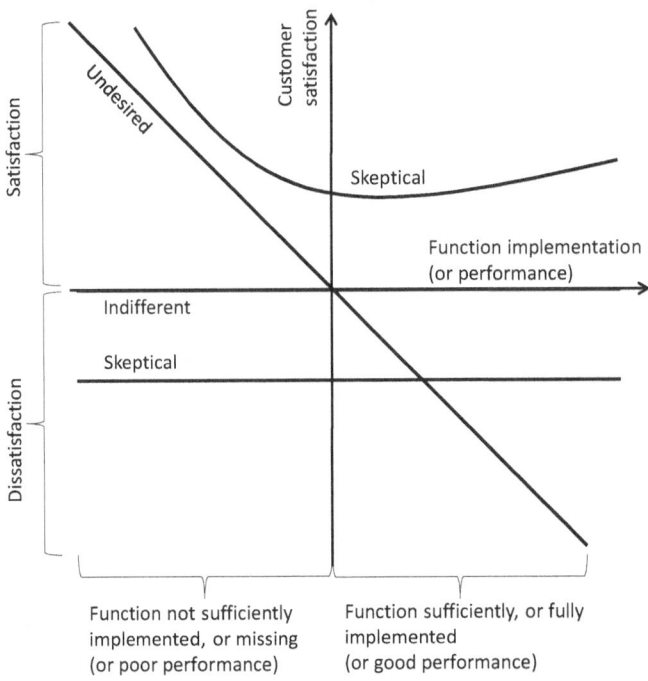

FIGURE 9.5 Representation of less common features ("undesired, "skeptical", and "indifferent") in the Kano diagram.

out to generate indifference, it is usually removed from the design. The "low fuel consumption" for a car can be indifferent for luxury cars owners, who do not care about fuel cost.

- "Undesired" (or "reverse") is a feature whose quality generates dissatisfaction. It is described by a curve with negative slope in the Kano diagram (i.e. the reverse of "desired" features). An "indifferent" feature can turn to "undesired" because of cost considerations. For instance, many customers do not care about casual clothes which last more than one season, since after one year they will be unfashionable. "Durability" tends thus to be "indifferent" for them. If, however they perceive that durability is associated with a higher cost, it could become "undesired": you do not want to pay an extra price for something you will not use. In the example of Figure 9.4 the lightness of a smartphone becomes undesired below 100 g.

9.3.4 Determine the Kano Category of a Product Feature

Now that we have realized that a product feature can be classified into a few main categories, the question is how to identify the category the feature belongs to. A reliable method is the so-called Kano survey, which uses inverse paired questions to ask the customer: "how do you feel if the function is provided" (or if a certain performance level is met)? and "how do you feel if this feature is not provided (or if a certain performance level is not met")? Standard answers like "very dissatisfied", "dissatisfied", "neutral", "satisfied", and "very satisfied" can be responded. Many combinations of answers are possible. From Table 9.3 we can see that, out of 25 combinations, 5 identify one of the three main idealized categories (cells marked in dark gray). This is further explained in Figure 9.6. Customers will, however, often provide blurred answers (cells marked in light gray). In the next section we will discuss how to average the answers of many customers and how to interpret the results. The cells corresponding to "skeptical" answers are typically reached only because of an error or a misunderstanding, which can be easily corrected by asking customers for further explanations.

Blurred answers are very common, since the categories outlined above are an idealization or reality and do not exactly fit it. Thus, you should resist the temptation to directly asking the customer questions like "is this feature for you expected,

TABLE 9.3
How Kano category is inferred from the answers to standardized questions

How do you feel if provided	How do you feel if not provided?				
	Very dissastisfied.	*Dissatisfied*	*Neutral*	*Satisfied*	*Very satisfied*
Very dissatisfied	Skept.	Skept	Undes.	Undes.	Undes.
Dissatisfied	Skept.	Skept.	Undes.	Undes.	Undes.
Neutral	Exp.	Exp./Des.	Ind.	Undes.	Undes.
Satisfied	Exp./Des.	Des.	Exc.	Skept.	Skept.
Very satisfied	Des.	Des./Exc.	Exc.	Skept.	Skept.

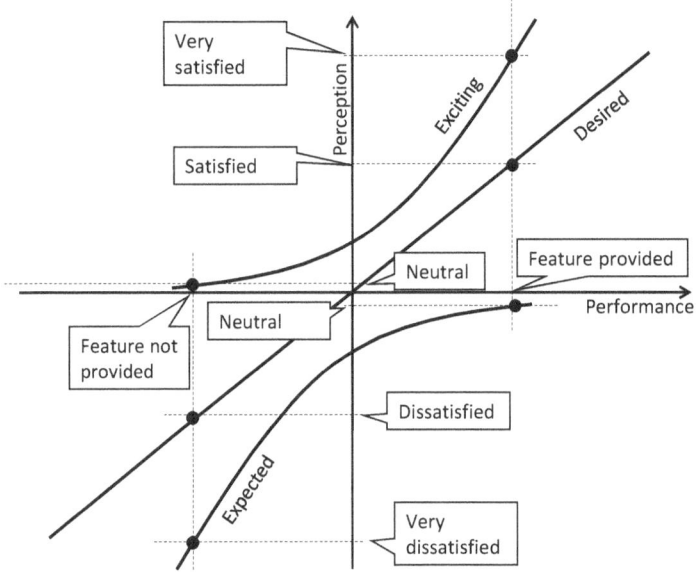

FIGURE 9.6   Relationship between answers to standardized questions and Kano category.

exciting or desired?" It could be challenging for the customers to answer, and their discomfort could easily generate misunderstandings and misleading answers.

9.4   PRACTICAL APPLICATION OF THE KANO METHOD

9.4.1   TERMINOLOGY

The terms used for the standardized answers should allow customers to express their feelings. Terms like "very dissatisfied" or "satisfied" appear logical once you know the Kano method. This does not mean that they are appropriate for all the customers: whether this simple set of terms is a good one, or what can be used instead, depends on the customer segment you are talking to, and it should be judged case by case. Before using the Kano method, you should work together with product experts and sales team of your company to create your set of terms.

For instance, the set discussed above turned out to be problematic for US customers in the study by Ross [4]. Similarly, a different terminology was necessary for international customers of a medium-sized enterprise manufacturing medical devices too. They needed to interview both native English speakers and customers who learned English as a foreign language, and created a different set of answers suitable to all of them (see Table 9.4).

Note that in Table 9.4 the answers to the question "how do you feel if the feature is provided?" are never negative, and the answers to the question "how do you feel if the feature is not provided" are never positive. Thus, this set does not allow identifying undesired features. This simplification was introduced because features designed as the solution to customer needs are very unlikely to be undesired.

TABLE 9.4
Replacement of standard answers with customized ones

Question	"Standard" Answers	Example of Alternative Answers
How do you feel if the feature is provided?	Very satisfied Satisfied Neutral	Wow, exciting! That's innovative… It would be useful I do not mind
How do you feel if the feature is not provided?	Neutral Dissatisfied Very dissatisfied	I do not mind It would be disappointing Without this the product is useless

9.4.2 IMPORTANCE OF FEATURES

Kano classification does not provide clear indications about the importance of a feature, except in the case of "expected" ones, which are logically very important. Since you will often have to drop some features contained in the proof of concept to save development time or to reduce manufacturing costs, it may be useful to have an additional parameter dedicated to importance. This information is partially redundant, and it helps to find gaps in customer reasoning and misunderstandings. To investigate importance, you can simply ask the customer "what problem or opportunity does this feature solve for you and how important is it that this be improved?" based on natural language, for instance, "extreme", "very strong", "strong", "moderate", and "weak". A weaker alternative, if you want to force an arbitration between features, is to sacrifice accuracy and ask the customer to assign a relative score to each feature.

9.4.3 CALCULATE MEAN VALUES

Once you have performed the Kano survey with many customers, you need to consolidate your results. As we have done with customer needs, the first step is to check the uniformity of judgment within the customer population, which is supposed to belong to the same customer segment. A substantial variability in the answers can indicate that you are dealing with sub-segments, and this discovery can lead to a deeper and very useful market understanding. For instance, Ross [4] reported that, in a study of automotive sector, a short automobile braking was perceived by men and women differently. Thus, the two sub-segments had to be evaluated separately. This led then to design two barking systems: one for family vehicles often purchased by women and one for performance cars often purchased by men.

If responses are homogeneous, you can simply take the most frequent evaluation (i.e. the mode) of Kano category (and possibly of importance). If responses are consistent with a single segment but are so "blurred" that you cannot clearly identify the mode, which is often the case, you need to perform some average. To do so, you need first to convert natural language to numbers. An example of possible scales is shown in Table 9.5. Here a geometrical progression is used for importance in agreement with the Weber–Fechner law. Answers from different customers are combined using

TABLE 9.5
Possible weights that can be associated with answers based on natural language

Question	Answer	Score	Mean
Importance	Extreme	1	Geometric
	Strong	0.5	
	Moderate	0.25	
	Weak	0.125	
Satisfaction if provided	Wow, exciting!	1	Arithmetic
	It would be useful	0.5	
	I do not mind	0	
Dissatisfaction if not provided	It is the bare minimum	1	Arithmetic
	It would be disappointing	0.5	
	Without this the product is useless	0	

geometric means. A pseudo-geometric progression, with "0" assigned to the lowest degree, is used for satisfaction and dissatisfaction. The reason is that the practical advantages offered by a geometrical progression with just 3 scale degrees are limited, while:(1) the results are much more intuitive if zero is used for a complete lack of satisfaction or dissatisfaction and (2) if necessary, you can use negative numbers and model undesired features. Consequently, arithmetic mean is used to average satisfaction judgments.

An example of data which could have been collected in the evaluation of regular Joe's design is shown in Table 9.6. To keep the example simple, just four customers are considered here, although this number would be insufficient in real situations. Six features are included in this example: four satisfy top customer needs (mobile phone interface and app, screens and consoles for back seats, move rear seats back, and increased wheelbase) and two, already provided by the competitors, turned out to be very important for the users (increased roof height and backup camera). For each feature, the team asked customers the importance first, then satisfaction if the feature is provided and dissatisfaction if the feature is not provided. If the customer answers that a feature has no importance, the questions about satisfaction and dissatisfaction are skipped, and the lowest score is assigned by default to these parameters.

9.4.4 DISPLAY RESULTS

A convenient way to display Kano results is a bubble plot. Every feature is represented by a bubble, whose radius is proportional to the importance. Its horizontal coordinate is dissatisfaction if the feature is not provided, and its vertical coordinate is satisfaction if it is provided. The plot area can be intuitively split into four quadrants, which will house exciting, not important, desired, and expected features (Figure 9.7). With the scales used in our example the "desired" feature quadrant is the largest because it must include the point (0.5, 0.5). Note that the bubble size should intuitively increase

TABLE 9.6
Example of Kano questionnaire involving four customers

Feature	Customer 1			Customer 2			Customer 3			Customer 4			Mean		
	R	Sat	Diss	R	Sat	Diss	R	Sat	Diss	R	Sat	Dis	R	Sat	Dis
Car–Mobile phone interface+app	1.00	1.00	0.00	0.50	0.50	0.00	1.00	1.00	0.50	1.00	1.00	0.00	0.84	0.88	0.13
Screens and console for back seat	0.50	0.50	0.50	0.50	1.00	0.00	0.50	1.00	0.50	0.50	0.50	0.50	0.50	0.75	0.38
Move rear seats back	0.25	0.50	0.50	0.50	0.50	0.50	0.50	0.50	0.50	0.25	0.50	0.50	0.35	0.50	0.50
Increased wheelbase	0.13	0.00	0.00	0.13	0.00	0.00	0.25	0.50	0.00	0.13	0.00	0.00	0.15	0.13	0.00
Increased roof height	0.25	0.50	0.00	0.25	0.00	0.50	0.13	0.00	0.00	0.50	0.50	0.00	0.25	0.25	0.13
Reversing camera	0.50	0.50	1.00	1.00	0.00	1.00	0.50	0.00	0.50	1.00	0.50	1.00	0.71	0.25	0.88

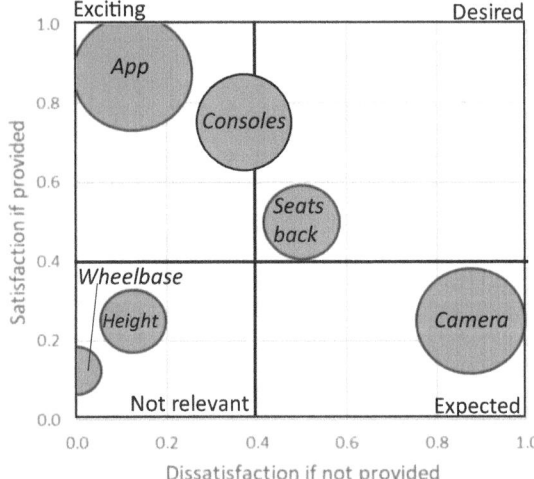

FIGURE 9.7 Representation of results for many features in the satisfaction/dissatisfaction plane.

with the distance from the origin; if it does not, for example, if a large bubble is in the "not important" quadrant, or if a small bubble is placed close to top edge of the plot, it is necessary to review the data to see if this uncommon behavior has a logical explanation, or if it is due to a mistake.

The plot in Figure 9.7 has been filled out with the exemplary data of Table 9.6. In this example, the logical outcome would be to:

• Reject the increase of roof height and wheelbase, since these features are not perceived as really important by the customers (although the design team thought they are good solutions to top needs).
• The shift of rear seats toward the back should be the minimum sufficient to be noticed by adults, and to provide a sufficient living space to the kids, but not more than that.
• The availability of a backup camera is very important to prevent strong dissatisfaction.
• At least one of the two exciting features should be implemented to provide excitement and differentiate the product from competition.

9.5   PRESENTATION TO INTERNAL STAKEHOLDERS

Once the evaluation with customers has been completed, and you think you have a sound high-level product design, it is time to present your results to internal stakeholders.

The product concept you are going to present should not be a complete surprise for them. Instead, you should have kept them constantly informed about the status of your investigation since the beginning. The feedback they provided during the project revealed their doubts, criticism, and what they are misunderstanding. Be sure that you address all these points in your final presentation so that its content will be aligned with the expectations of the audience. The worst thing which can happen is that you talk about a subject different from what managers want to hear. You do not have to agree on every comment you have received, but you should show that it has been taken into consideration. It is also a good habit to share your presentation a few days in advance so that you will not have to discuss all details, and more time will be dedicated to questions and discussion.

Do not forget that every group of people communicates in a different way. The language you used with customers was not the same you used with engineers. Similarly, the language you use now must be targeted to internal stakeholders. The data are always the same, but the way they are structured and presented must be tailored to the listeners.

Typically, the presentation can be organized in three blocks, explaining (1) why you should do something (in view of the needs of your company and customers), (2) what you propose to do, and (3) how can this be achieved (Figure 9.8). More in detail:

1. Why does the status quo offer you an opportunity?
   - The business goal you are pursuing.
   - The customer segmentation: clearly identify one or a few segments which are your target. If other segments are generating substantial profit with existing products, explain why they have been excluded.
   - The life of customers in this segment today: what are they doing and with which product.
   - The main pains of the customer.

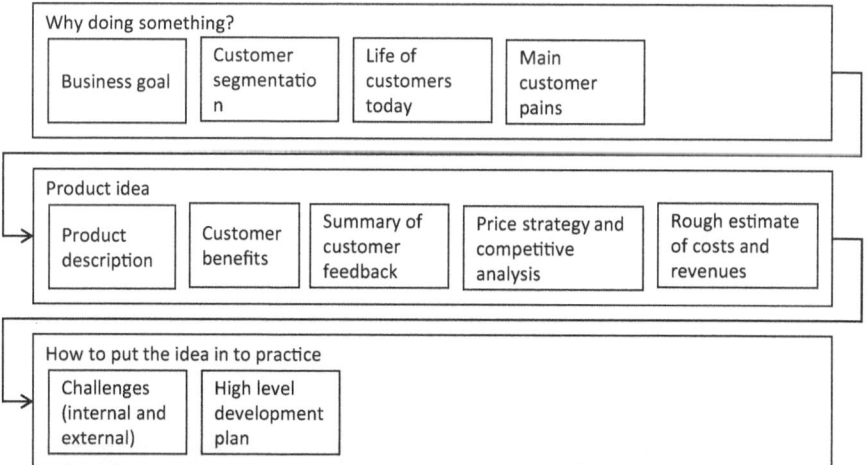

FIGURE 9.8   Possible structure for a presentation to internal stakeholders.

2.   What is your product idea?
   *   A description of the product you are proposing, including the benefit it provides to the customers. You can substantially use what you showed to the customers during the proof of concept.
   *   A summary of customer feedback you collected during the proof of concept. This feedback must show that the product is really perceived by the customers as a solution to their problems, for which they would pay. A nice Kano plot will probably excite your audience and will trigger a fruitful discussion.
   *   The pricing strategy and the competitive analysis.
   *   A very high level estimate of revenues, costs, and profits. This will constitute the seed of the future business case.
3.   How the idea can be put into practice?
   *   The challenges you expect because of internal and external factors, as well as factors that could be beneficial. This is often structured as a strengths, weaknesses, opportunities, and threats (SWOT) table. If you decide to use such a format, be aware that a SWOT table tends to become overfilled with text, which makes it of little use. Whatever format you use, be sure to point out the most important facts and leave out what is not essential.
   *   A high-level development plan, with a rough estimate of development costs, resources, and time. This will bridge project management once the development starts. Be sure that the next steps and the most urgent decisions are highlighted.

It is difficult for most people to remain focused and attentive to subjects for more than 40 minutes. Thus, your presentation should not take more than 30 minutes. This will leave enough time for discussion. This time constraint forces you to focus on what is essential, and to omit what is not. At this stage, you need to tell a story which can be easily followed and makes sense. Do not hesitate to trade off accuracy for intuitiveness and clarity. If the project is approved, you will have better opportunities to show a rigorous data analysis.

NOTE

1   https://en.wikipedia.org/wiki/Comparison_of_smartphones#2004

BIBLIOGRAPHY

[1]   Renvoisé, R., Morin, C., Neuromarketing, Thomas Nelson Publishing, 2007.
[2]   Dixon, M., Brent, A., The challenger sale: How to take control of the customer conversation, Penguin, February 2013.
[3]   Kano, N., Seraku, N., Takahashi, F., Tsuji, S., "Attractive quality and must-be quality", Journal of the Japanese Society for Quality Control, vol. 41, pp. 39–48, 1984.
[4]   Ross, H. M. "The new Kano model - how to really excite your customers", presented at the 26th Symposium on Quality Function Deployment, Charleston (SC), December 5, 2014.

10 Beyond Product Design

Although this book is focused on product design, the basic principles it outlines can be applied with great benefit in all the stages of a product's lifetime. These principles are therefore reviewed and discussed in this section, together with some tools that could be very helpful to put the basic principles into practice in the continuation of product development.

10.1 BASIC PRINCIPLES

"Principles" are usually taught in courses of the QFD institute, and are discussed in ISO16355 too, as the philosophical foundation of QFD. The list presented here has been adapted from these sources to target the audience and scope of this book.

10.1.1 Focus with Priority

Focus is required for the best use of limited resources and it is very beneficial at many stages of product design. Examples of focusing we have encountered are setting of project goals (what does the company really want?), customer segment targeting (which customers bring more value?), customer needs (which are the most important needs of customers in the most important segment?), and design selection (which is the most efficient, scalable, marketable, or profitable way to satisfy these needs?).

Unfortunately, there is a fundamental psychological hurdle associated with focusing and setting priorities: you must implicitly accept putting something else to the side. And deciding what "not to do" is a tough management decision, much more difficult than deciding what "to do". Techniques to identify sound priorities with minimum bias are necessary and help, but they are not enough. Besides that, one needs qualities like leadership and accountability which can be exerted only if the company culture encourages them. If this is not the case, anyone will avoid as much as possible taking clear decisions. This is the reason why in QFD methods this culture, if not present, is created with the direct involvement of top management. At a minimum, try to apply this principle within your own work group, or at least to your personal duties, and focus your efforts on what generates the highest value.

DOI: 10.1201/9781003544845-10

10.1.2   LISTEN TO THE VOICE OF CUSTOMERS

Be humble and accept the fact that, despite how deep your product experience may be, your understanding of customers is always an assumption that diverges over time. You must listen to them in order to be sure you understand their needs, and you must listen with the proper attitude: being able to question your own beliefs and being ready to change them if evidence requires it. Pretending customers will buy what you think they need is a strategy with very low chance of success.

Product design is not the only time at which the voice of customers is collected: remarkable examples are prototype testing and feedback of customers who purchased the product. During the development, you should try to regularly show prototypes to customers to be sure you are moving in the right direction (paying of course due attention to avoid premature leaking of your plan to competitors). The way in which this is performed and the timing depend on the product and the industry: in early development stages you often rely on demonstrations made to customers by the development team; during the prosecution of product development, you tend to move toward independent customer testing, possibly at their site.

The easiest products to be tested are probably software (SW) cloud applications for a wide public: customers will just need an internet connection. Besides that, the cost of the test for the company is very limited. Tests become more demanding when moving to service and hardware products, especially those sold in small volumes. However, the basic principle remains always valid: only customers can tell you if what you are doing is right, and periodic tests must be included in your development plan. It would be very dangerous to limit your scope to the tests which are mandatory for your product because of regulations, or to those made by reference customers for marketing reasons.

If many customers can independently test the product, which is a very positive situation, you will need to collect standardized feedback from them. In this case, differently from what we have suggested in the rest of the book, you can use a written survey. The reason is that after a customer has touched the product with their hands, they can judge it without risk of misunderstanding. Ensure, however, that questions are properly chosen (less is always better) and formulated. Use closed-ended questions whenever possible, and meaningful predefined answers. You do not have to discuss Kano categories anymore, but for different features it may be worth to inquire the quality of implementation and the importance of the feature for the customer. We can consider as a real-life example a study about an SW tool for industry. An SW prototype was shown to several customers, who were asked to judge 16 features in terms of two parameters: importance for the user and quality of implementation. For both parameters predefined answers based on natural language were proposed, which were then translated into weights in a geometric progression during analysis (weights are shown in Table 10.1). The responses from different customers were then averaged using geometric mean. The final results are shown in Figure 10.1. Although no product feature exceeds customer expectation, the implementation appears good, since almost all the important features are implemented at least in a "fully satisfactory" way. It is probably necessary to improve further feature F6, which would be

TABLE 10.1

Parameters used to judge the features of an SW prototype

Parameter	Natural Language Judgment	Weight
Importance for the user	Essential	1
	Very useful	0.5
	Nice to have	0.25
	Almost useless	0.125
uality of implementation	Fully satisfactory	1
	Just OK	0.5
	Helpful but not enough	0.25
	Not usable	0.125

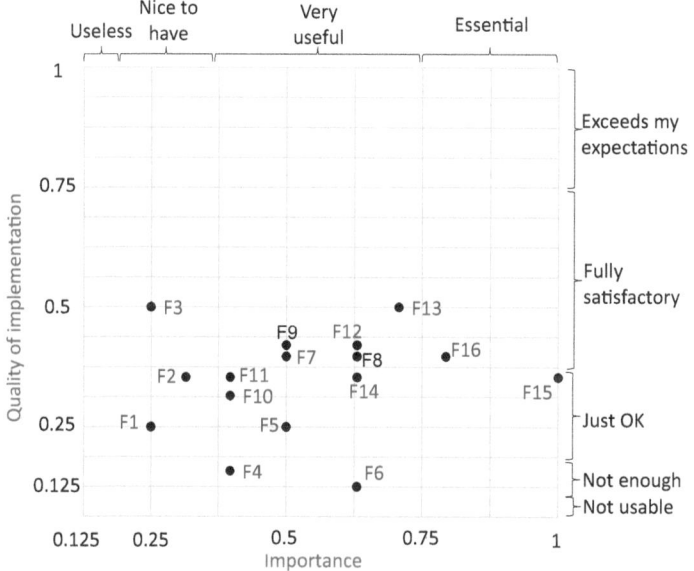

FIGURE 10.1 Real-life example of evaluation of features of an SW tool for industry. Importance (independently from implementation) and quality of implementation (independently from importance) are evaluated for 16 product features.

"very useful" but its implementation is "not enough". If resources are available, it may be worth improving F15, which is deemed as "essential", but whose implementation is "just OK".

Similarly, the feedback of customers who purchased the product should be carefully analyzed to identify future improvements. For instance, feedback from a customer who purchased the product but is not using it could let you understand why many do not even place an order.

The extent to which the voice of customers can be used throughout the product lifecycle heavily depends on the way you organize your projects. Most projects can be seen as the mixture of two idealized, extreme cases:

1. The sequential model (also called "waterfall model"), which stresses the importance of initial planning, with a rigid, monodirectional flow of requirements: decisions taken at a certain stage are requirements for the next stage. Sequential-like models can be preferred in industries such as construction industry, where product changes are very expensive, and many efforts are spent preparing an accurate plan to be executed.
2. The iterative model, which puts emphasis on outcome improvement through refinement cycles, with back-and-forth communication between actors in different stages. A very popular example of an iterative model is Agile software development.

There are a few reasons according to which iterative processes best foster a customer-centric spirit and allow a systematic use of the voice of customers. First, you can make good use of the voice of customers only if you can make changes as you learn more. Remember the adage "no plan survives contact with the enemy", which in our context means that only customers can tell you if your creation is the right product for them. They may adore features which you put to the side and dislike those that you think are fundamental. Second, if you want to address needs which are not yet solved by existing products, you must exploit the creativity of your team. And the most effective way to suppress creativity is to force people to follow rigid instructions.

10.1.3   Understand the Situation

We saw how this understanding is important when we discussed the translation of the voice of customers into customer needs by considering all available information, which is not limited to what the customer says, but extends to the observation of what the customer does, and the feelings and emotions they let you perceive. The validity of this principle extends beyond design: all data you receive must be understood based on the context if you want to avoid misconceptions and generate false information. A statement can have a very different meaning depending on who made it, to which scope, and in which circumstances, and it is possible that the true meaning cannot be understood without additional investigation.

10.1.4   Understand Causality

You must be in control of your actions and decisions regarding products. If you know what you are doing and why, you can offer quality on purpose. Even if at some point you fail, you will have better chances to understand what went wrong and take corrective actions in product updates or next generation releases. This is possible if you understand the causality relationships which link together the different stages of product development, commercialization, and even aftersales support. We have already discussed how this principle applies to the relationships between the voice of customers and customer needs, as well as between customer needs and high-level functional requirements. The same understanding of causality (i.e. traceability) should be present in all phases of the product lifecycle.

As a matter of fact, the design team must ensure that they generate a set of engineering characteristics which can be translated into a real product in view of many constraints: know-how (what the company knows or what can be learned), technology (what is possible with current state of technology), cost (limited by what the customer is willing to pay in view of perceived value and competition), and acceptable time to market (the company needs a return on investment within a certain period; besides that, customer needs may be solved by competition if you are too slow).

Traceability is particularly necessary to make the link between customer needs and product lifecycle tasks explicit to everybody. We must stress that all employees will have to continuously take decisions on their own to accomplish their tasks. The more these decisions are aligned to the final goal (the satisfaction of top customer need), the higher the quality of final result. It is important that this alignment involves sales and marketing, who in the end communicate to the customers and handle their objections. This is explained in Figure 10.2. In this example the product must solve three main customer needs (gray cells). Two product developments are considered: a "bad" one (top) and a "good" one (bottom). White circles indicate which customer need is addressed by the efforts of different departments. In the bad development, initial requirements are aligned to customer needs, but alignment is lost during the process because developers do not have a clear awareness of these needs. In the final

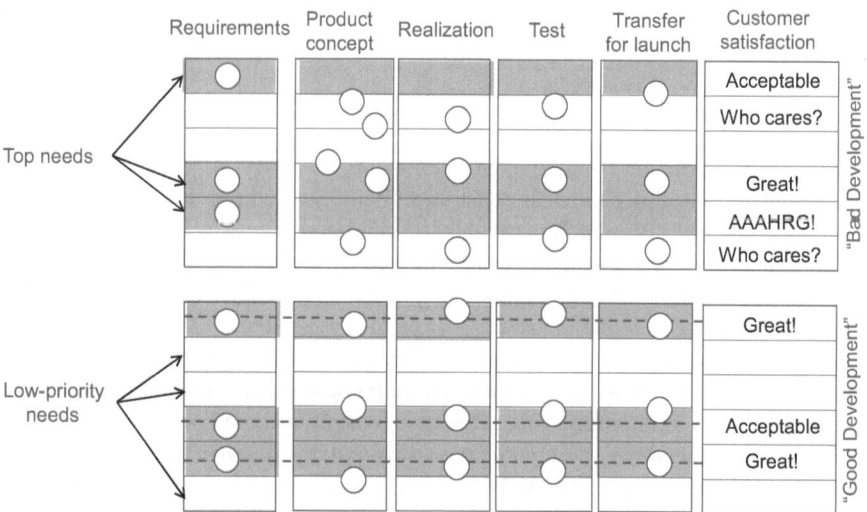

FIGURE 10.2 Sketch of how top needs (gray cells) are addressed at different stages of product development by departmental efforts (circles) in the case of a "bad" (top) and a "good" product development.

product of this example only one need is fully solved, one is partially solved, and one is completely missed. On the other hand, resources have been wasted to provide a feature addressing a need which was not really important to solve. In this situation customer value may be achieved only by chance.

In the "good development", the staff is aware of the customer needs which must be solved; thus, they are able to take the right decisions and adjust the direction of their work, and despite unavoidable quality fluctuation, the final product satisfies the three top needs. As a result, high value is created on purpose.

10.1.5   FAVOR MARKET-IN THINKING AGAINST PRODUCT-OUT ATTITUDE

According to the market-in approach (also called customer-centered or market-oriented approach), the goal of product development is to bring to the market a benefit the customer is willing to pay for. The product is just the means to provide this benefit. This counterintuitive approach, which deeply permeates this book, collides often with the more engineer-intuitive product-out approach (also called technology-driven or product-oriented approach), in which the goal is to release a product. A market-in approach is driven by customer needs; product-out is driven by a technical understanding of what the company can do. In other words, the fundamental question of market-in is "what will customers buy?" opposed to the question "what can we make?" of product-out.

The product-out attitude can be convenient to improve a product which is already successful so that you may be able to assume with less risk that it satisfies important customer needs. It could be equally a product in your portfolio, or a successful product of the competition. The market-in approach is much more suited to the development of new products, especially if they are supposed to be innovative. A product-out approach to a new product may represent a risky leap in the dark.

10.1.6   IMPROVE COMMUNICATION

Product development and commercialization is teamwork, and a team can work only if there is efficient communication among its members. A major cornerstone of communication in this context is "customer focus", which provides a North Star to the cooperation of different departments. As Shigeru Mizuno, a founder of classical QFD, pointed out, product quality cannot be reached if different groups within a company are focused on reaching only their own goals ([1], p. 353). Thus, customer focus must override departmental focus. In QFD methods, it is a task of upper management to avoid a conflict between customer focus and performance goals of departments. If customer-centric methods are embraced by individuals on their own, they will have to exert, or develop, persuasion skills to create a consensus about how to solve customer needs. The bidirectional communication fostered by iterative models (discussed below) could help reach this goal. Well-stated customer needs are also useful to have the voice of customers heard throughout the full product development and commercialization processes, as discussed in Section 10.1.4.

A useful technique to promote iterative communication is the so-called catch-ball. Catch-ball is commonly used in Hoshin Kanri, a method for enterprise strategic planning, to create an agreement between managers at different hierarchy levels about objectives: the person in the higher position explains their expectations and needs to their direct-reports in the next lower position, who generates a proposal to achieve these needs. The two will discuss it through several iterations until a consensus is reached and a solution acceptable for both is identified. In our context this method can be applied to different steps in product development. In particular, the description of requirements, especially nonfunctional ones, does not have to be exhaustive, but focused on the key information needed by developers to take decisions about details on their own.

Consider again the example of the car for the "regular Joe-like" customer. We have entertainment displays in the rear seats, where illumination can change from darkness to bright sunlight depending on time, traveling direction, car location, etc. The development team decided that, to allow a comfortable view under different conditions, display brightness should automatically change according to the environment. John and Mary are two engineers: Mary is an expert in sensors, and she should select an ambient light sensor. She has knowledge about electronics, but not as much as John, who is the expert in charge of developing the electronics which will bring the sensor information to the display control system. Mary's decisions can now heavily influence John's creativity and the quality of the outcomes. Consider, for instance, these two situations:

1. Rigid decisions: Mary gives John the datasheet of the best sensor she has selected. Besides that, she gives him rigid technical requirements for the electronics, including indications about the architecture he must use.
2. "Catch-ball" behavior: Mary shows John her shortlist of three possible sensors, indicating the one she believes is the most suited to their challenge. She invites him to create one or more design proposals around this sensor. Mary gives John the freedom to create alternatives using the other sensors in the shortlist too. The two agree to meet again one week later to select together the best solution.

The first option would probably just frustrate John, who would develop no dedication to the results of his job and will perform tasks without much passion (such a phenomenon is known as "reification" in philosophy). The second option can exploit John's creativity and creates an opportunity to detect errors on both sides. John will enjoy his freedom and generate more data, and this is not likely to cost more time, since a passionate person is much more efficient than a discouraged one.

10.1.7 STRIVE TO BE SYSTEMATIC

In QFD it is recommended to define a process for implementing the customer-centric techniques in the company to enjoy the many benefits of standardization. Among others, we can mention reaching a consistent level of quality, do not reinvent the wheel, facilitate review of results, build company expertise, and generalize lessons learned across product families and divisions. If you apply the customer-centric principles on your own, standardization attempts could be excessive. Nevertheless, it is worthwhile at least to structure your data, take decisions based on evidence, and document the decision-making processes (as it will be discussed in a while, a key prerequisite for good decision-making is the ability to perform fair comparisons). This attitude will make easier the creation of consensus in the team, and the review of the work whenever necessary. Further, it will make explicit to the team the implicit knowledge of its members. If you have the chance to use the same method in a second project, or in several projects, use the workflow you used in the previous one as the starting point so that you will build expertise with it.

10.2   FAIR JUDGMENT

10.2.1   CORNERSTONES

Fair judgment is a key element in understanding causality, improving communication within the team, and taking the right decisions. In particular, good decision-making is essential to create the awareness and the consensus necessary to convince team members that a solution is the right one for the customers and to have them working together with commitment and passion.

The cornerstones of fair judgments are the following:

1. Never judge a single element but consider alternatives. There is always at least one alternative even if you have the impression that you have just one option: what will happen if you do nothing.
2. Be sure the elements you are going to judge really represent alternatives: in other words, they must be "homogeneous" (that is "commensurable").
3. Clearly state the criterion or criteria used to evaluate the alternatives, and their relative importance.
4. Evaluate all alternatives against all criteria. If you have listed real alternatives, they can all be evaluated against all criteria. If this is not the case, the items you called "alternative" are not really alternatives.

We can review these recommendations with an example. Let's assume that Anne wants to organize a party, and she asks if you think that apples should be served as a fruit. This question is insufficient to perform a fair judgment; let's see how you can help Anne in her decision process. Anne should first clarify which are the alternatives to apples. Can she consider other fruits? Or a cake instead? Anne answers that the party celebrates local products, and the only alternative to apples are citrus fruits. Now you have more options, which are not however alternatives: apples are a very specific fruit, but citrus fruits are a family which contains very different fruits like oranges, lemons, or grapefruits. Which one shall we consider? Anne clarifies that the only citrus fruit she can imagine at the party are oranges because others are too sour. This is a helpful statement since, despite proverbs, you can easily compare apples and oranges. You just need to clarify the criteria used for the judgment. Anne explains that her criteria are "seasonality" and "the possibility to eat it standing up". She adds: "I will go for apples since we can easily eat them standing up and oranges are out of season". At this point, you recognize that she is not evaluating both alternatives against all criteria, and you let her notice that this judgment is unfair: oranges can be easily peeled and eaten with hands, maybe better than apples, which could be troublesome for people who do not like the taste of their skin. And the party is taking place in June in Italy, so apples are out of season anyhow. Your last effort is to let Anne evaluate again both fruits against all her criteria (local, seasonal, easy to eat standing, and taste) before taking a final, fair decision whatever it is. Anne could also consider the relative importance of her criteria.

10.2.2 RELATIVE AND ABSOLUTE CHANGES

Humans are much more skilled in performing relative judgments ("comparisons") than absolute judgments ("measurements"). As a matter of fact, we have developed instruments to perform measurements for us. Many examples are possible, for instance:

- It is easy to figure out who is the tallest between two people standing one next to the other, but it is very difficult to estimate the height of one person without a yardstick, unless this person is very close to you, in which case you can compare with yourself. This task is particularly difficult if you see the person in a photo against a neutral uniform background, without objects in the vicinity.
- Play in sequence two keys of a piano. Almost everybody can tell which of the two pitches is the highest. However, very few people can hear a single pitch and tell which piano key was pressed to generate it, even if they have the possibility to actually play the keys.

The reason for our deficit in absolute judgment is that it involves the relationship between a single stimulus and some information held in our short-term memory [2]. A measurement instrument is built in such a way to "remember" its calibration, but for humans it is much more difficult to recollect past perception. The number of accurate discriminations that can be made in this way is extremely limited, for example, 6 musical pitches, 5 levels of loudness, and 10–15 positions on a line. On the other hand, comparative judgment is based on the relationship between two stimuli, both present at the same time, and it allows extremely fine discriminations.

To further clarify the different performance which we can reach with absolute and relative judgments, look at Figure 10.3. Pay attention to the brightness of the

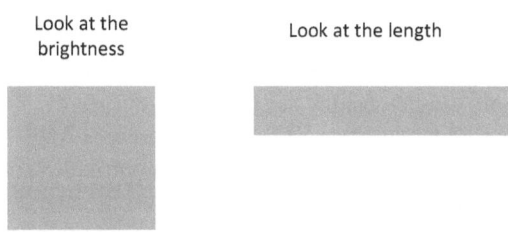

FIGURE 10.3 Example to show the difficulty of absolute judgment.

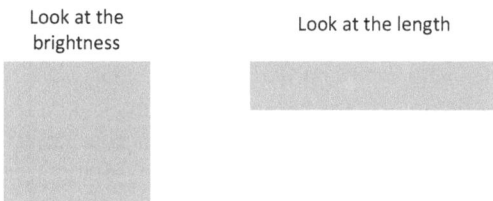

Look at the
brightness

Look at the length

FIGURE 10.4   Same geometrical shapes of Figure 10.3, but with a slight (4%) decrease in brightness (left) and length (right).

square, and the length of the rectangle. Then go to the end of this chapter and look at Figure 10.4. You will find there a square with slightly different brightness, and a rectangle with slightly different length. Can you tell which square is the brightest? Or which rectangle is the longest? Probably not: we are not good instruments to measure absolute lengths and brightness. Now look at the same shapes, placed side by side at the end of the book (Figure 13.2): when looking at this picture you can probably easily tell which square is the brightest and which rectangle the longest, although the difference is just as large as 4%.

Our ability in relative judgments is the reason why we have stressed so much the necessity that customers prioritize customer needs only when they are compared one against the other in a hierarchy diagram.

10.2.3   REVIEW OF DECISION TOOLS

In the previous sections we have introduced several tools to compare alternatives. The most remarkable ones are:

- The affinity and the hierarchy diagrams, to make sense of unstructured data and aggregate them into groups of commensurable elements. We have used them widely to classify customer needs of different levels of generality.
- The AHP pairwise comparison, to compare alternatives against a single criterion. It was used to quantitatively evaluate customer needs with the same level of generality based on their "priority" for the customer.
- The decision matrix (also called "Pugh method") to compare alternatives using multiple parameters. It was used, for instance, to compare second level requirements versus their suitability to implement first level requirements.

These tools were combined when we introduced the house of quality, where product functional requirements were compared using customer needs prioritized with AHP. An example which recapitulates this concept is shown in Table 10.2. In this example needs from a hierarchy diagram are prioritized with AHP. Their global priorities are used as weights in the decision matrix. The contribution of every functional requirement to satisfy customer needs is evaluated using natural language. Grades of natural language are then translated into numbers using a logarithmic stimulus–response law: grades "extreme", "very strong", "strong", "moderate", and "weak" are

TABLE 10.2
Example of house of quality

	Needs		1st level requirements			
	No.	Global priority	Req. A	Req. B	Req. C	Req. D
Need 1 (0.3)	Need 1	0.3	Extreme	Weak	Weak	Very Strong
Need 2.1 (0.4)	Need 2.1	0.28	Mod.	Strong	Very Strong	Weak
Need 2.2 (0.6)	Need 2.2	0.42	Weak	Very Strong	Strong	Extreme
Need 2 (0.7)						

0.7 ·0.6

	Req. A	Req. B	Req. C	Req. D	
$1·0.3 + 0.125·0.28 + 0.063·0.42$	0.36	0.30	0.26	0.59	Priority (raw)
$0.36 / (0.36+0.30+0.26+0.59)$	0.24	0.20	0.17	0.39	Priority (normalized)

Note: The AHP process used to prioritize customer needs is schematically represented on the left-hand side. Calculation details are shown in grey boxes.

replaced by 1, 0.5, 0.25, 0.125, and 0.063, respectively. These numbers are multiplied by needs weights to calculate the functional requirement priorities.

10.2.4 SUPER PUGH METHOD

The house of quality method can be adapted to virtually any decision process you may encounter in design and deployment. An interesting generalization is constituted by the "Super Pugh" process, which is very useful to compare alternatives of both an objective technical nature and a fuzzy subjective nature, for instance, technologies suitable to implement a feature. Super Pugh mathematics is very similar to the house of quality one, with an important preliminary step to select decision criteria. This procedure can be described by the following few steps:

- List the pros and cons of every alternative. This is easy because this is what most people instinctively do. This information is useful, as long as it is understood that it cannot be directly used to carry out a comparison. For example, "small size" should be neutralized to "size."
- Use pros and cons of alternatives to create an affinity diagram, and rationalize it by adding missing parts, thus generating a hierarchy diagram.
- Prioritize the elements in the hierarchy diagram with AHP.
- Evaluate your alternative against the most important elements of the hierarchy diagram using a decision matrix. In some cases, it may be preferred to evaluate all alternatives before prioritizing the criteria to reduce bias in the results. In this case criteria weights are included in the calculation as a last step.

As an example, let's assume that you must decide the size of the displays to entertain the kids in the car designed for the regular Joe customer. An engineer in your team, who is fond of infotainment systems and car customization, takes charge of shortlisting the most promising options. It is made clear from the beginning that these options cannot exceed a preliminary budget limit, and that only one movie/videogame player will be available in the car. The engineer comes to the next meeting with few slides in which he presents three alternatives:

1. Two "small size" displays, in the range 13"–15", to be attached to the back of the front seats. They are cheap, but they do not look very exciting, since they are just a bit larger than common tablets which kids could carry on.
2. Two "medium-size" displays, in the range 17"–19". They are clearly larger than any tablet, but only a little more exciting since they have the size of gaming laptop screens. Besides that, they are more expensive than smaller displays.
3. One "large" display in the range 22"–24". It is the size of a large desktop monitor and would look huge inside a car. It is not cheap, but one display would be large enough for everybody sitting in the back. However, it will have to be mounted in the center of the car, either beyond the center console or attached to the ceiling. Both options will create mounting challenges.

The team arranges these elements in a hierarchy diagram and evaluates them with AHP. Results are shown in Figure 10.5. Cost savings and excitement were present in the initial data. However, as often is the case, new elements (single user usability and multiple user usability) have been added during the construction of the hierarchy diagram. "User perception" received a much higher score than "cost savings" because the cost of all the proposed alternatives would be acceptable. The "usability (multiple users)" considers the possibility of playful interaction between two and three kids enjoying the same movie or videogame together.

This AHP required the construction of three matrices for pairwise comparison: one 2×2 matrix to compare "cost saving" versus "user perception"; one 2×2 matrix to compare "display" vs "installation"; and one 3×3 matrix to compare "initial excitement",

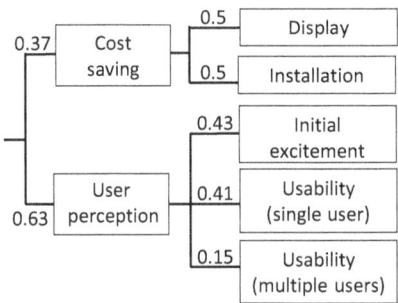

FIGURE 10.5 Hierarchy diagram with the AHP evaluation of decision criteria for the car displays. Note that the priorities are relative priorities within the groups, and they sum up to one (or 100% if you prefer).

TABLE 10.3
Matrix used to prioritize the three needs related to user perception

$(1 \cdot 3 \cdot 1 \cdot 3 \cdot 1/5)^{1/5}$

	Initial excitement	Usability (single user)	Usability (multiple users)
Initial excitement	1	1.125	2.667
Usability (single user)	0.889	1	2.954
Usability (multiple users)	0.375	0.339	1

1/2.954

Note: Calculation details are explained in gray boxes.

TABLE 10.4
Quantitative decision matrix used to prioritize display size versus top decision criteria in the car infotainment system example

Criterion		Display configuration			
Statement	Weight	2x(13"-15")	2x(17"-19")	1x(22"-24")	
Display (saving)	0.18	0.57	0.07	0.76	
Installation (saving)	0.18	0.76	0.50	0.13	
Initial excitement	0.27	0.16	0.33	0.57	
Usability (single user)	0.26	0.19	0.50	0.76	

Strong
Very strong
Very strong
Extreme
Extreme

$(0.5 \cdot 0.25 \cdot 1 \cdot 0.5 \cdot 1)^{1/5}$

0.63 · 0.41

0.18 · 0.57 +
0.18 · 0.76 +
0.27 · 0.16 +
0.26 · 0.19

0.34	0.33	0.52	Priority (row)
0.279	0.28	0.44	Priority (normalized)

0.34 / (0.34 + 0.33 + 0.52)

Note: Calculation details are shown in gray areas.

"usability (single user)", and "usability (multiple users)". As an example, the 3×3 matrix is shown in Table 10.3. Prioritization is performed by a team of five people; thus, the three independent values of the matrix (1.125, 2.667, and 2.954) are obtained as the geometric average of five judgments. The leading eigenvalue of this matrix is 3.005: the fact that it is very close to 3 indicates that the judgments are, on average, consistent (note that averaging may have reduced the impact of possible inconsistency of individual scores). The components of the normalized leading eigenvector are (0.43; 0.41; 0.15), indicating that "initial excitement" and "usability (single user)" have close importance and are both much more important than "usability (multiple users)".

The prioritization of display alternatives against the decision criteria is shown in Table 10.4. The criterion "Usability (multiple users)" has been excluded because its absolute weight (0.63·0.15 = 0.1) is much smaller than the weights of the two other higher priority criteria. In this example, relative judgments were used, and all judgments were performed by a team of five people. Thus, the entries in the decision matrix are the geometric average of their scores, which were expressed using natural language. For instance, in the case of the cell highlighted in the figure ("0.57"), the members judged the importance of a large display to provide initial excitement as "very strong", "strong", "extreme", "very strong", and "extreme", respectively, which were translated into 0.5, 0.25, 1, 0.5, and 1, respectively, using the usual logarithmic stimulus–response law. In the end, a single, large display receives the highest score because it is outstanding in view of all criteria except the installation cost, which however was not the most important criterion.

In the example of Table 10.4 the average of subjective evaluations is used for all criteria. Depending on circumstances, it could be worth to use a different method:

TABLE 10.5
Example of quantitative prioritization of display size versus savings

Display size	2x(13"-15")	2x(17"-19")	1x(22"-24")
Price	$495	$675	$360
Priority (method 1)	0.27	0.00	0.47
Priority (method 2)	0.49	0.20	0.71

(675-495)/675   (800-360)/(800-180)

Note: Two possible prioritization methods are considered.

- For critical subjective criteria you may want to prioritize options with a pairwise comparison instead of providing direct scores. This will increase accuracy. As an alternative you can ask the evaluation of a panel of experts who are acquainted to judge a certain parameter. For instance, in the case of usability a panel of experts could evaluate usability of every option using a system usability scale (SUS), which provides a score of 0–100 (and which you can normalize to 1). Note that asking a score from an expert is much safer than asking a score from a generic customer or colleague, if they are trained to provide judgment and minimize subjectivity.
- For objective criteria, you can use real data. For instance, if you know the typical price of displays of different sizes, you can use them in your calculation. It is important that you properly normalize these numbers so that you could get a meaningful dimensionless value. There is not a universal rule to calculate priorities from quantitative data, and you need to choose the calculation procedure which is the most suitable for your case. Two examples with our display savings are shown in Table 10.5. The easiest way to prioritize the three savings is to (1) calculate the saving with respect to the most expensive option and (2) normalize the saving to the latter. This calculation is indicated as "method 1" in the table. Another possibility is to define the highest acceptable price (which should correspond to a priority 0) and the ideal price you could hope to pay if you spend enough time to search the market and bargain with suppliers (which should correspond to a priority 1). Using $800 and $180 for these two extreme prices, you get the calculation indicated as "method 2" in the table.

It is of course possible to use the super Pugh procedure for a qualitative evaluation too. An example is given in Table 10.6. If well conducted, a qualitative evaluation can reveal important insights, and provide a fair comparison as well, although a quantitative one has the potential to provide higher accuracy if you can input reliable data. Whether you should use a qualitative evaluation or collect the data necessary to perform a quantitative evaluation depends on the accuracy you need. Typically, the need for accuracy increases with progression in the project. Management decisions require less accuracy than engineering decisions and the tools must be selected accordingly.

TABLE 10.6

Qualitative decision matrix used to prioritize display size versus top decision criteria in the car infotainment system example

Criterion	Display Size		
	2x(13″–15″)	2x(17″–19″)	1x(22″–24″)
Display (saving)	☺	☹	☺
Installation (saving)	☺☺	☺	☹☹
Initial excitement	☺	☺	☺☺
Usability (single user)	☺	☺☺	☺☺
Usability (multiple users)	☺	☺	☺☺

BIBLIOGRAPHY

[1] Mizuno N., Epilogue-Interview with Shigeru Mizuno," in *QFD: The customer-driven approach to quality planning and deployment*, edited by Mizuno S. and Akao Y, Asian Productivity Organization, Tokio (Japan), p. 353, 1994.
[2] Blumenthal A. L., The process of cognition, Prentice-Hall, Englewood Cliffs (NJ), 1977.

11   Cope with Time Pressure

A quick drafting of product design is a common request in contemporary industry. To reduce design time, you can try to optimize everything which is under your control. However, you will find soon that there is a lower bound to the duration of the design process, dictated by the time necessary to interact with customers. The reason is that you cannot control the availability of customers and their speed of response.

Let's try to make our reasoning more quantitative. Interaction with many customers is concentrated in three stages:

1. Collection of the voice of customers to discover customer needs.
2. Needs prioritization.
3. Proof of concept.

You interact with customers to create the hierarchy diagram too, but this happens only once and does not dominate the bill of efforts. Such one-occasion interactions are neglected here. We assume that the efforts spent to go through stages $i = 1 - 3$ are proportional to the number of involved customers:

$$Efforts = k_1 \cdot N_1 + k_2 \cdot N_2 + k_3 \cdot N_3, \qquad (11.1)$$

where N_i is the number of customers involved in step i, and k_i is a proportionality constant. The number of customers in the second stage N_2 is typically the largest, but as a first approximation, we assume that the number of customers involved in the three stages are the same ($N_1 \approx N_2 \approx N_3$) and that k_i constants are close to the common value k:

$$Efforts \approx 3k \cdot N, \qquad (11.2)$$

where N and k depend on the product. The design of brand-new products in an unexplored market segment will require more time per customer than the replacement of an existing one. We consider as a reference case the design draft for a new product dedicated to a well-known application, which has already been covered by other products in your portfolio. This endeavor would probably require about 6 months and

DOI: 10.1201/9781003544845-11

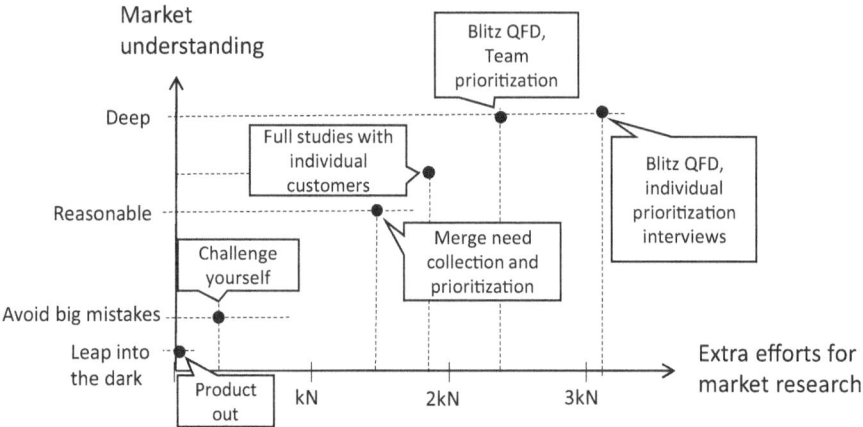

FIGURE 11.1 Sketch of possible simplified methods compared to the one outlined in previous sections ("Blitz QFD, individual prioritization interviews").

TABLE 11.1
Summary of alternative customer-centric design methods, with the efforts they required (versus Blitz QFD) and major drawbacks

Method	Relative efforts	Limitations
Blitz QFD, individual prioritization interviews	100%	-
Blitz QFD, team prioritization	–80%	Some bias, but only if customers in the panel are competitors.
Full studies with individual customers	–70%	Customers are not exposed to needs of peers.
Merge needs collection and prioritization	–60%	Needs prioritization is biased by your paradigms. Unspoken needs are lost.
Challenge yourself	–15%	Limited market insights.

$N = 15$ customers, which means $k \approx 180/45 = 4$ calendar days. This gives an idea of the cost of adding a customer to the project.

Depending on the context of your work, this time could be available or not. Luckily, if you are put under time pressure, there are ways to shorten the market analysis and generate quicker a high-level product design. Of course, there is always a price to pay to complete your work faster, and this price is typically a significant loss of accuracy since accuracy requires time. However, you always need to compare alternatives to take decisions, which in this case are (1) the design without market research and (2) the design with the market research you can afford in both time and budget. A market analysis with limited accuracy is typically much better than no analysis at all. Therefore, if you have no other choice, it is worth trading time for accuracy.

An overview of speedup possibilities is given in Figure 11.1. Close to the vertical axis there is a purely product-out approach: it represents the "cheapest" approach because no extra efforts are spent to understand the needs of the customers. At the other end of the vertical scale, we have the method described in previous sections, which is based on "Blitz QFD", requiring efforts in the order of 3 kN. Alternatives requiring effort in between these two extremities are presented in the following sections. Additional details about these design methods are given in Table 11.1, where the main limitations of each one are also indicated.

11.1   BLITZ QFD WITH TEAM PRIORITIZATION

Needs prioritization can be made more efficient by gathering the customers together and performing the prioritization in one single workshop. The organization of such a meeting will not be easy and it will probably require additional expenses, for instance, to support customers' travel and accommodation costs, but it is likely to require much less time than N multiple interviews. Consequently, your efforts will be somewhere between 2 kN and 3 kN. In our reference example, you could complete the study in about 4 months. It is important to note that this method is the only one that allows you to save time without losing accuracy.

Depending on circumstances, team prioritization may provide even better results than the one based on multiple interviews. This is because broader knowledge becomes available as a result of the panel debate. A special caution is, however, needed if customers are competitors of each other because in this case their group evaluations may be biased. The recommended procedure to moderate a team prioritization is as follows:

- Let customers verbally express their ideas about which need is the most important and why. The customers should talk in a sequence from the less authoritative to the most authoritative (for instance, in view of experience, fame, or hierarchy position). This minimizes the influence of the most authoritative customers on the others.
- Let customers provide a judgment in the reverse sequence (the first customer to talk will be the last one to vote). If the same customer were the first to talk and to vote, his influence on the others would be exaggerated. If judgments are provided with natural language, be prepared to convert them into numerical values on the fly. If the respondents are of multiple languages, it may be preferred to use the numerical judgment instead of natural language.
- If you use AHP, skip a branch of the hierarchy diagram only if the need is not important to satisfy for all the customers. If a need is not the top on the average but is the most important to satisfy for some customers, investigate its branch during a private discussion with them after the end of the meeting. The additional information you collect will help you to verify the homogeneity of your customer segment. Also be sure that your AHP SW or spreadsheet can

accommodate multiple judgments and calculate their geometric average and the consistency ratio of every group of needs.
- Ask a colleague to join you to take note of the comments made by customers: they are valuable information.
- Discuss the results with the team of customers before closing the meeting and be sure that you have a clear understanding of the reasons behind the judgments.

11.2   FULL STUDIES WITH INDIVIDUAL CUSTOMERS

This simplified method was developed for the company Sandvik Coromant, a cutting tools producer [1, 2]. This company already had an established method to identify customer needs. However, they had to visit 20–25 customers worldwide for every market research, and they needed a new solution to dramatically shorten the time to market of their products. The foundation of their new method was that the investigation team spends a significant time at customer sites (typically 2 days per customer) to:

- Discuss the customer process model (CPM) with the customer and perform the Gemba visit.
- Analyze the voice of the customer and identify customer needs.
- Review the needs with the customer and give them a structure.
- Prioritize the needs of the specific customer.

With this method the order of magnitude of efforts would be about 2 kN. Sandvik reported that it represented a reduction of one half with respect to their previous procedures.

The main drawback of this method is that a customer is not exposed to the needs of other customers. Additionally, there is no guarantee that different customers will disclose the same needs. However, it worked fine for Sandvik, since they observed that all the customers worldwide typically shared three out of their five top needs, the other being region and industry specific.

11.3   MERGE COLLECTION AND PRIORITIZATION OF CUSTOMER NEEDS

This simplified method, which has been widely used at IBA Dosimetry, is sketched in Figure 11.2. A first set of relatively unstructured interviews and Gemba visits are carried out in parallel to the study of available literature to understand the customer context, to create a sketch of the CPM and the skeleton of the hierarchy diagram. Note that since literature plays now such an important role, it must be validated and updated with customers because they may be aware of important documents which, although public, are not known to your team.

A second set of interviews are then used to refine the hierarchy diagram and score the needs. This set will usually contain more customers than the first one: the needs are typically the same for all customers in the segment, but their importance is more customer specific, and a larger population sample size is necessary to reduce

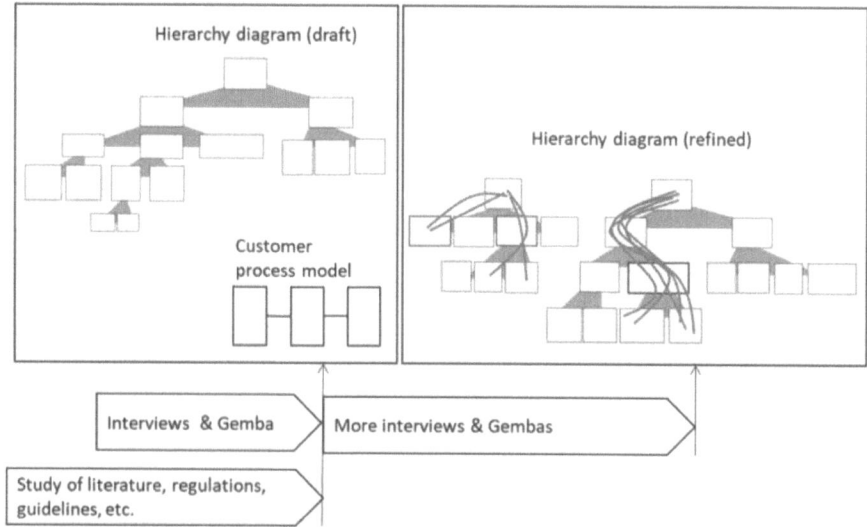

FIGURE 11.2 Sketch of the method to merge the stages of customer needs collection and prioritization. The tasks carried out using this method are shown by the bars at the bottom of the picture. The content of the squared boxes on top illustrates the state of your knowledge when these tasks are accomplished.

statistical variation. This second stage is more standardized, and for each customer, it includes a few steps:

1. Look at the CPM with the customer and identify pains, opportunities, and needs.
2. Review the hierarchy diagram. Focus on levels that are not too abstract and go through all needs at this level. Double-check that you and the customer have the same understanding of these needs.
3. Ask the customer if they see some missed important needs in the diagram, and if the set is consistent with the CPM review you performed together. If a need from the CPM review cannot be identified in the diagram, it is added on the fly.
4. Ask the customer to score the needs. Since at this point the residual interview time is probably not that long, and since this method is not aimed at the highest accuracy, a scoring method simpler than pairwise comparison may be used. Once the top need has been identified, double-check that the assignment is correct by asking "why is it so important?" Be sure you clearly understand the "why": this is your most important take-away. It is not infrequent that a customer gives the highest score to a need proposed not by them but by another customer. This is evidence once more of the power of the hierarchy diagram.
5. Rationalize: after the interview, carefully review the diagram. It is possible that, after changes on the fly with the customer, it is not very "clean"

anymore: if necessary, simplify branches and make it more intuitive so that you can clearly explain it to the next customer. It is essential that you are comfortable with the diagram so that you can explain it in a convincing way and answer possible customer questions. Answers like "I am not sure about this need, but it was added because of another customer" are not confidence-building.

With this method you can expect a complexity close to 2 kN. For our reference project it would mean as few as 3 months of work. This method offers an additional benefit too: you have had since the beginning a picture of the market. This picture is initially blurred, but it becomes clearer and clearer with time. Thus, if you must report internally the status of your investigation, you can always present something tangible, which is highly appreciated by management. You can also decide to stop the investigation earlier, or extend its duration, depending on time, cost, competitive pressures, and available resources.

These benefits are not free. They are associated with two significant drawbacks:

a. As already mentioned, you lose accuracy. In particular, you introduce a framing bias since customers are confronted with your assumptions and the needs of the customers contacted previously. If you talk to a large enough number of customers, your results will converge toward an "average top need" because the statistics levels out fluctuations. However, the "average top need" you discover could be somehow different from the true top need.

b. You may overlook needs. Important unspoken needs are not always present, but if they are, they can be uncovered only by looking at the customer in a neutral way, without being limited by your paradigms.

In view of drawback (b) this method must be used with judgment. Give the hidden complexity of reality the opportunity to surprise you!

11.4 CHALLENGE YOURSELF

The minimum you can do to gain insights about customer needs is to challenge yourself. This method can be very useful to gently introduce customer-centric principles in a company with a product-out mindset: colleagues will appreciate the benefits of customer centricity without feeling threatened by an abrupt change.

This means that if you are already dealing with a product idea generated by a product-out approach, you should try to figure out what is the benefit for the customers it would deliver and how large this benefit would be. Are substantial problems solved which cannot be solved in other ways? Is the life or work of the customer really improving? Try to put these reflections in writing, and validate them with a few customers.

Efforts here are <<kN. In our reference example this can probably be achieved in less than 1 month. Even if you will not improve your market understanding a lot, you have at least the chance to rule out major mistakes. Remember the $80,000price

monstrosity designed in the Simpsons cartoon by the company which blindly followed Homer's indications. If you go this way, it may be worth remembering the basic principles outlined in Chapter 10 and try to constantly apply them.

BIBLIOGRAPHY

[1] Bylund, N., Wolf, M., Mazur, G., "Reducing lead time in cutting tool development by implementing Blitz QFD", Proceedings of ICED 09, the 17th International Conference on Engineering Design, vol. 6, pt. 2, Palo Alto (CA), pp. 255–266, 2009.
[2] Mazur Glenn, H., Bylund, N., "Globalizing Gemba Visits for Multinationals," Transactions from the 15st symposium on quality function deployment, Savannah (GA), December 5, 2009, pp. 61–78.

12 Do Not Let Your Mind Deceive You

Human beings are the result of a long evolution. The species *Homo sapiens* emerged around 300,000 years ago. Thereafter our ancestors lived for a long time as hunter-gatherers; only about 10,000 years ago they started to practice agriculture, to domesticate animals, and to settle down, in a process which led to the development of civilizations. This means that our mind is optimized to survive within small groups in a wild environment, not to design a product for a company which operates in a complex technological society with billions of members.

This is particularly true for the mental process we indicate as intuition: the ability to acquire knowledge without recourse to conscious reasoning. Being able to take quick decisions based on a limited amount of data can be very important in certain circumstances. However, intuition tends to be inaccurate and suffers from cognitive errors (fallacies in the rational process) and affective errors (when one's emotional state influences decision-making). These errors, or biases, are considered responsible for many diagnosis errors in medicine and have been widely studied (see Refs. [1], [2], [3], and [4], which are generally valid for Sections 12.2–12.6).

Deliberation is likely to outperform intuition when one has conscious access to sufficient data [5]. However, deliberation is not free from risks either. It develops according to patterns which, depending on the context, are indicated with different names: models, theories, mindsets, dogma, or paradigms. They provide a foundation of presumed truth upon which thoughts can be efficiently built. At the same time, they delimitate the space where thoughts can evolve. In many cases paradigms can help to quickly solve a problem, but in others they can prevent you from reaching a solution or even seeing the problem itself.

Although we cannot easily change the way our minds work, we can become aware of how it works. This awareness, together with the applications of the "debiasing techniques" proposed in the following, can help to overcome limitations in our reasoning and to disclose the true logical connection between customer pains, customer needs, and product requirements.

DOI: 10.1201/9781003544845-12

12.1   BE AWARE THAT YOU FOLLOW PARADIGMS

The term "paradigm" was popularized by the treatise "the structure of scientific revolutions" [6], a study in the field of history of science which proposes that successive transitions from one paradigm to another via revolution are the usual way science develops. According to this study, when scientists try to understand a new phenomenon, they move through several phases (see Figure 12.1):

1.   Pre-paradigm phase: no consensus on any theory.
2.   Normal science: a paradigm becomes dominant, and problems are solved within it. In this phase knowledge increases by accumulation.
3.   Crisis: anomalies are revealed which cannot be solved within the dominant paradigm. At this point the only way to progress is to recognize that the old paradigm is no longer suited to describe reality, to abandon it, and to create a new one ("paradigm shift").

As an example, we can retrospectively describe with a paradigm shift the "Copernican revolution" [7]. Ptolemy's geocentric system was developed during last two centuries before Christ and was the paradigm for astronomy till the beginning of the modern age. It postulates that the earth is a sphere in the center of the universe. This turned out to be a very good scientific model, which was successful in describing

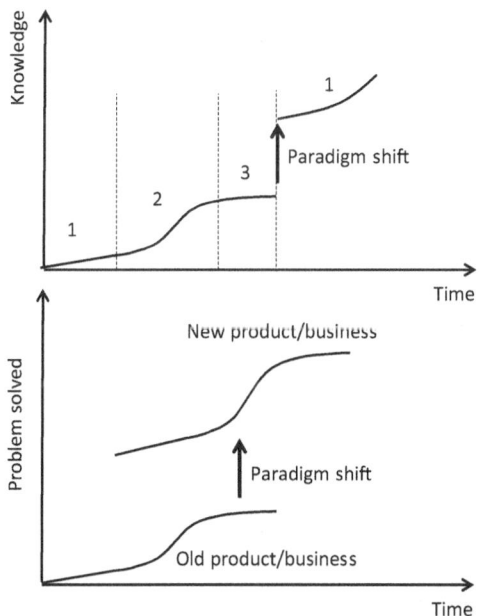

FIGURE 12.1   Top sketch: scientific knowledge evolution according to Kuhn's theory of paradigm shift. Bottom sketch: how driving a paradigm shift can lead to a competitive advantage.

observations of the motion of the sun, planets, and other celestial bodies, which seem to rotate around us. As time passed, more and more experimental observations were performed, and with increasing accuracy. It became difficult to explain them in the framework of the geocentric model, which entered a crisis. From the 15th to the 17th century, a new paradigm (the heliocentric system where the planets revolved around the sun) was developed by Copernicus, Galileo, Kepler, and Newton. In the 18th century, the paradigm shift was completed. Consider that a heliocentric idea was already available in Ptolemy's age, but it was not widely adopted at that time because, without Galileo's telescopic observations and Newton's theory of gravity, it did not provide clear advantages over the geocentric model.

The theory of Kuhn has been widely criticized (see, e.g., [8]). For instance, it is not difficult to identify retrospectively many paradigm shifts in the history of science, like Dalton's atomic theory, Darwin's natural selection, germ theory of diseases, quantum mechanics, relativity, etc. However, this does not automatically imply that these shifts occurred in a revolutionary way during the course of time, nor that this applies to most science and not just to selected cases. It is not our scope to support or refuse Kuhn's theses. What is important in this context is to observe that it is worth to critically reviewing your beliefs: although in many cases paradigms offer you a safe framework to work within, in other it blinds you (as well as your competitors) to new opportunities.

Similarly to science, one can interpret many changes in technology and business as paradigm shifts, and possible examples become more and more numerous in recent times [9]. A few shifts recently revolutionized our way of living, like the rise of the internet, smartphones, social networks, the cloud, and at-home working. Others lead new technologies to completely replace old ones (e.g. flat screen television sets, LED lights, and digital cameras). From the observation of these selected examples, one can infer that:

a.   Every paradigm uncovers problems it cannot solve. It sets the stage for a new paradigm, but it does not disclose the new rules.
b.   In most cases a solution is already available, somewhere.
c.   A paradigm shift can create a competitive advantage: as sketched in Figure 12.1 (bottom), if you successfully drive a paradigm shift, your product will solve customers' problems much better than all other existing products (or in other words, it will provide a much higher benefit).

The verb "can" in statement (c) is used on purpose. Practically, "driving a paradigm shift" means to take the risk of making huge investments in innovative products. If you fail, it will be to the benefit of competitors who will learn from your errors: to be or not to be a pioneer is a complicated dilemma.

Of course, it is not easy to challenge our own beliefs, otherwise they would not carry this name. There are, however, techniques which can facilitate this operation. For instance:

• Discuss with someone who knows little or nothing about your business. Such a person is the most likely to change your paradigm.

- Ask yourself: "what changes could put your organization back to zero?"
- Ask a customer: "which are the problems that should be solved, and you do not have a clue of how to do it?" or "what is impossible to do in your field but, if done, would fundamentally change things for the better?"

12.2 LOOK FOR A PARTNER

A partner can be very helpful in the process of collecting customer feedback. The most obvious reason is that it is difficult for a single person to interact with a customer and take notes at the same time. Besides that, a partner can also provide feedback about the quality of conversation with the customer, which can be biased in many ways. In particular, he can help you to avoid two common biases: the illusion of attention and the framing effect.

The illusion of attention occurs when an individual fails to perceive an unexpected stimulus that is in plain sight [5]. This typically happens when the mind is very busy finding something else. When it becomes impossible for one to attend to all the stimuli in each situation, a temporary blindness effect can take place as a result, and you do not see what you are looking at. This is one of the reasons why it is difficult to find something you do not know and generate innovative ideas.[1]

The "framing effect" is represented by the fact that answers are influenced by the way a question is framed. Consider as an example the following three questions a father could make to his teenage son:

a. What do you wish for your next birthday?
b. Do you want a motorcycle for your next birthday?
c. You don't want a motorcycle for your next birthday, do you?

The first one is very open and neutral: one can expect any response. The second is already providing a price range for the present the kid can expect to receive and, at the same time, provides an option. The last one clearly brings the hope that the kid forgets the "dangerous" idea of riding a motorcycle. Option (b) could come from a person who is already a motorcycle rider, and (c) from someone scared by the idea of being. Depending on the personality of the kid, questions (b) and (c) can influence his decision in one direction or in the other.

The framing effect is heavily used deliberately in advertising, media, and politics. It is not rare that two newspapers or networks report the same fact with different words so that it is in one case very negative, in the other very positive. And would you prefer to put your savings in an investment which turned out to be loss-making 30% of the time, or one offering the 70% of chance to make a profit?

If your questions to customers are framed and not asked in a neutral way, there is the risk that you will hear simply what you want to hear. The result could be that you develop the product you like, not the one solving customer needs. A partner can effectively point out a framing in many cases and help you to improve communication with customers. Another strategy to mitigate the framing effect is to prefer questions which encourage the customers to talk freely about their concerns rather than respond to your inquiries.

12.3   MONITOR YOUR EMOTIONS

Emotions have the power to make life worth living. At the same time, they have the power to make a market analysis useless. This could be the case if you let your emotional state influence your decisions (the so-called affective error). It could be induced, for instance, by the environment (e.g. stress and fatigue), by the influence of personal relationships, or by your mood. You may dislike a customer, but this is not a good reason to neglect their feedback, or to treat it with superficiality: they may be a good sample of an important customer segment.

Additionally, we have the tendency to explain others' behavior by putting emphasis on their internal characteristics, rather than external factors, a cognitive error known as the "fundamental attribution error". Typical examples are thoughts that almost any car driver has from time to time, especially those living in big cities. Consider, for instance, these sentences, which you probably already heard: "Look, how this idiot parked the car!" or "if these idiots would just take the bus, I could be home now, instead of being stuck in a traffic jam". It may be that the "idiot's car" is improperly parked, and for sure you are never alone in a traffic jam, but it is very unlikely that this is due to the intelligence of other drivers. It may even be that we are idiots for driving during rush hour! Other factors like the lack of public transportation, roadblocks, or even the weather are much more plausible causes.

As an additional example, consider a customer who uses your product in the wrong way and criticizes it: it is tempting to say that "he is incompetent". However, it may be that the customer is very competent, but the product's documentation is wrong, or that the use of the product is not intuitive. Similarly, a customer could have chosen a product from the competition, and praised it, although you have objective reasons to think that your product outperforms it. Is he "a customer without a clue?" Or does your competitor, which has better marketing highlight the benefits of their product, hiding any drawbacks? Or is your product overengineered?

It is interesting to note that when looking at our own failures, we usually have the opposite tendency and overemphasize external factors: "it was not because of me...".

Because of these errors, emotions should be kept out of our product design process or at least we should be able to separate our internal inferences from the voice of customers. To understand if your judgment is biased, just ask yourself: do I have negative feelings about this subject? Am I using stereotypes? If the answer is "yes", it is likely that your reasoning is biased, and you should discuss your findings with a colleague before going ahead. A partner is helpful against this error too.

12.4   TAKE NOTES

There are many good reasons to take notes and decrease the reliance on memory. The most obvious one is that we cannot remember everything we perceive, and much information would be simply forgotten if not quickly recorded. This fact is widely accepted and for this reason we discussed many methods to record data in Chapter 4. There is however another fact which we tend to ignore: our recollection can be distorted, or just a creation of fantasy.

This is called the "illusion of memory". Chabris and Simons [5] proposed a simple test to highlight this illusion. To take the test you can (1) slowly read the list of 15 words in the footnote and turn the page,[2] (2) wait a few minutes, and (3) try to write down as many as you can remember. You can try before going ahead: your results will be very interesting.

As you can imagine, most people cannot remember all the words in the list. However, what is more surprising is that typically part of the "recollected" words which are written down are created by the writer! If you compare the list you wrote down with the original one, you are likely to find new words, probably "sleep". The reason is that the content of memory depends both on what happened and on how we made sense of what happened, which includes rationalizing the information and finding connections between its parts. All the words in the list were about sleep, thus "sleep" is easily added by the mind to create a coherent piece of information which can be easily remembered.

Note that the vividness of our recollections is tied to how they affect us emotionally, and that vividness is usually used as an indicator of accuracy. Thus, one should beware of memories accompanied by strong emotions and vivid details: they are just as likely to be wrong as others, but you are far less likely to realize it.

David, one of the authors, remembers that many years ago he was lying in his bed at night while reading Bram Stocker's "Dracula". At the point when the vampire struggles to enter a bedroom through a closed window, a gust of wind opened David's window wide, letting him literally jump out of his bed. This is a very vivid recollection. The funny thing is that one of his closest friends claims to have experienced the same event. This probably happened to one of them, who told it to friends. The other was impressed and created a copy of this story in his memory. Now we have two people with the same memory, and no way to say who jumped out of his bed!

When you compare your memories about customer feedback with the notes taken during interviews or Gemba visits, you will find differences. Be sure that you annotate all important things: if you find something in your memory which is not in the notes, you will know it has been created by your imagination.

12.5 STATE THE BOUNDARIES OF YOUR DATA

We treat confidence as a signal of a person's skill and knowledge. It is a common habit to dislike a physician who shows limited confidence, hesitates to make a diagnosis, and prescribes additional analyses. In these circumstances it is tempting to conclude that "he does not know what he is doing". However: is this behavior an indication of a lack of knowledge, or self-awareness? To what extent can we be sure that a prompt diagnosis from a confident doctor is correct? He may be a victim of the very common cognitive errors known as overconfidence: the tendency to believe we know more than we do.

Toilets provide a nice example of overconfidence [5]. Many people think they know how a toilet works, but they do not, and this can be easily seen by asking them for an explanation. All they really understand is how to work a toilet. They mistake their knowledge of what happens (a pressure difference, sucking most of the water out of the bowl, is generated when the water rushes quickly through the bowl and completely fills up the siphon) for an understanding of why it happens (the water in the bowl is flushed when you press the handle).

We can conclude of course that confidence is not a good reason to give a higher weight to the voice of a customer. But customer overconfidence is just a minor risk. Everybody wants to be confident, in view of its social reward, and we want to be confident too. Thus, we easily end up pretending we know more than what we do. It is very common to replace what a customer said with what we suppose a customer meant. It would be fair to report "customer A told me the fact B, and I think this is because of the reason C"; unfortunately, faults in our mental processes often lead us to simplify it into "All customers are reporting evidence C". This behavior is largely understandable: you want to appear smart, and to be sure to convey to management a simple and captivating message. On the other hand, you should find a way to foster your reputation and deliver sharp messages without distorting reality.

An effective way to fight overconfidence is to state the boundaries of the data you are using. How many customers did you visit? Where? From which customer segment? "12 customers in UK" says much more than "a lot of customers". This kind of information can be easily summarized in one concise table or in intuitive pictures which convey a lot of information (in our example it could be a UK map with flags on it). It will show you are not only smart, but self-aware too and, most of all, you will help colleagues to identify possible blind spots in your study.

12.6   DOCUMENT YOUR DECISION-MAKING PROCESS

We have the tendency to conclude too fast our decision-making processes. This cognitive error is named "premature close" and is a major cause of diagnostic errors in medicine. When we rush toward a premature close, our decision is dictated by our intuition and results are likely to be inaccurate and possibly totally wrong. This is because we have evolved to:

- Identify patterns. Pattern perception is central to our lives and mostly very useful. In many cases it helps us to make sense of our perception, as in the example of 15 words related to sleep.
- Explain the patterns by looking at events which happen together (correlation) and see if one is the cause of the other (causality).

If these two processes are not performed by investing the due time and effort, we can easily see patterns in randomness and mix up causality and correlation. This is probably because, in some circumstances, a wrong understanding of causality is better than no understanding at all. Let's assume that you live in the wild, and you find at sunrise the corpse of a friend devoured by a beast in a canyon. A few weeks later you find another companion in the same condition. You correlate being out at night with being devoured and conclude that the risk of encountering a beast is higher during the night. As a result, you remain hidden in your lair from sunset to sunrise. This conclusion could be dramatically wrong. Maybe the beast eats only corpses and being out in darkness merely increases your likelihood to stumble, fall in the creek, and die. You misunderstood correlation for causality, but nevertheless your wrong understanding saves your life.

Unfortunately, in the context of product design these cognitive errors can bring more troubles than benefits. You want to find true patterns in the voice of customers and establish causality relationship: the product causes the satisfaction of customer needs; satisfaction of customer needs causes the relief from pain.

First, you must avoid finding patterns in randomness. If you wear a talisman and nothing bad happens to you for a few days, you may be tempted to infer that the talisman protects you, which is of course a wrong conclusion. Even worse: if during the third day something bad happens, you could conclude that "well, it is not a perfect one, but still a good talisman". The third day will be promptly forgotten, but you may remember very well that nothing bad happened during the fourth day. This is called "confirmational bias", or "cherry picking": it is the tendency to look for confirming evidence to support your theory, rather than look for disconfirming evidence to refute it. Most conspiracy theories are based on detecting patterns in events that seem to help us to understand why they happened. The more you believe your theory, the more likely you are to fall prey to this illusion.

Second, you must understand that, although correlation may imply causation, it is not always the case. This misunderstanding can be due to several reasons. For instance, it may be difficult to distinguish the cause and the effect. Consider the correlation between recreational drug use and psychiatric disorders: many think that the drugs cause the disorders, but it could also be that people with a preexisting psychiatric condition have a higher tendency to use drugs looking for relief. Another case in which causation can be easily wrongly interpreted is when the correlating events have a common cause. A popular and exaggerated example is the correlation between ice cream consumption and the rate of drowning deaths: both increase in summer. Of course, consuming ice cream does not increase your likelihood of drowning, but in summer you are more likely to both eat ice cream and swim: the warm weather is the common cause of both events.

A documentation of your decision-making process is an effective measure against cognitive errors like premature close and confirmational bias. When documenting the decision process, you must specify which data you are using and which not, which may reveal if you deliberately favored data confirming your hypothesis, or that you have ignored data against it. Additionally, you must explicitly state why there is a cause–effect relationship. As soon as you start writing down your reasoning, you will find many missing connections and blank spots, which will require further reflections. Moreover, this written documentation will allow others to review your deductions.

The connections leading from the context through data to the conclusion can be documented with simple tools like a few written statements or diagrams like those used for root cause analysis. Consider the following example in which the description of a decision-making process evolves from a first, unsatisfactory draft to a mature version:

- The coffee is terrible, I must buy a new coffee machine.
- I used the same machine and the same coffee beans for 5 years. The coffee always tasted good. Now it is terrible. I must buy a new coffee machine.

- I used the same machine and the same coffee beans for 5 years. The coffee always tasted good. Now it is terrible. Friends using the same beans do not report any change in quality. I must buy a new coffee machine.
- I used the same machine and the same coffee beans for 5 years. The coffee always tasted good. Now it is terrible. Friends using the same beans have not reported any change in quality. The user manual suggests cleaning the machine every 6 months. I never cleaned the machine during its lifetime. I shall clean it. If it does not help, I must replace my coffee machine.

The first statement is an intuition dictated by emotions. The correlation between coffee taste and machine age sounds reasonable. It is possible that an old machine, in certain cases, brews bad coffee. However, it is also clear that in this statement there is not enough data to support this explanation. The last sentence is the result of a much more careful reflection. It has been recognized that alternative causes for bad coffee exist, and that further investigations are needed. This deliberation could let you save hundreds of dollars by investing a few minutes in cleaning the coffee machine.

If you have documented your decision-making processes, you do not need to say "trust me" anymore. If your analysis is reasonably good, there is no reason for you to claim trust: you can just explain the reasons behind your proposal by showing your decision-making process. If you are not able to explain your conclusions, they could be equally right or wrong. This does not mean that your personal experience has no value: it does, but you need to understand its limits: "I have had many coffee machines in my life. Every time the taste of coffee worsened, I bought a new one, and the taste always improved". Your experience shows that buying a new machine is likely to be a solution to the problem but shows that you could have wasted a lot of many too.

12.7 RECAPITULATION AND FINAL REMARKS

Figure 12.2 summarizes, in a very simplified way, how at different stages of product design process the debiasing techniques discussed in this section can help to neutralize the effect of cognitive errors and blindness due to paradigms, to finalize an innovative design.

There is one last illusion that is worth discussion: the illusion of potential [5]. It lets us think that our mind hosts unexploited mental ability. The illusion combines several beliefs: that we have the potential to perform at much higher levels than what we currently do; that this potential can be exploited with simple methods; that we have been so lucky to find an exploitation method that most people ignore.

This illusion is behind the success of most books about investment and stocks trading which, in turn, are great examples of cheery picking and wrong causality: the fact that an investment strategy worked for one person in a certain context does not mean that it will work for everybody. Maybe thousands of people tried the same, bad strategy, and the only one who by chance succeeded wrote a book in the end.

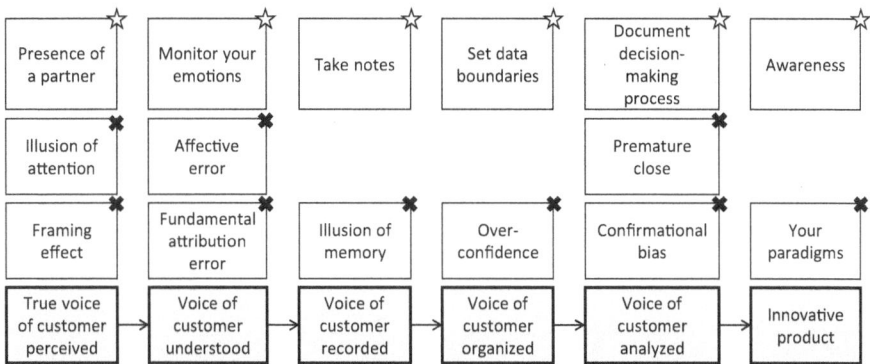

FIGURE 12.2 Sketch showing how, at different stages of design process (boxes with thick edge and gray filling), the proposed debiasing techniques (boxes with white filling marked with stars) can neutralize the effect of biases (boxes with white filling marked with crosses).

The illusion of potential shall not let you misunderstand the value of the customer-centric design method. It is not a recipe, or an equation, which accepts your professional problems as input and generates a faultless, innovative product design as an output. This method provides a set of tools, mostly based on common sense, which were developed and collected in the framework of the QFD methodology. According to the direct experience of the authors, these tools can help you to collect good market data and make sense of them. In the end, however, the quality of your thinking and the quality of your data will determine the quality of your results.

NOTES

1   This is effectively explained by the first video at www.theinvisiblegorilla.com, which summarizes the results of a famous psychology experiment. We recommend that you look at it before continuing reading these pages.
2   Bed, rest, awake, tired, dream, wake, snooze, blanket, doze, slumber, snore, nap, peace, yawn, drowsy.

BIBLIOGRAPHY

[1]   Croskerry, P., "The importance of cognitive errors in diagnosis and strategies to minimize them", Academic Medicine, vol. 78, p. 775, 2003.
[2]   Croskerry, P., Abbass, A., Wu, A. W., "Emotional influences in patient safety", Journal of Patient Safety, vol. 6, no. 4, pp. 199–205, 2010.
[3]   Groopman, J., How doctors think, Mariner Books, 2007.
[4]   Saposnik, G. et al., "Cognitive biases associated with medical decisions: a systematic review", BMC Medical Informatics and Decision Making, vol. 16, p. 138, 2016.
[5]   Chabris, C., Simons, D., The invisible gorilla, Harper Collins Publisher, 2010.
[6]   Kuhn, T. K., The structure of scientific revolutions, the University of Chicago Press, 1962.

[7] Kuhn, T. K., The Copernican revolution, Harvard University Press, Cambridge (MA), 1957.
[8] Toulmin, S., Human understanding, Princeton University Press, Princeton (NJ), 1972.
[9] Barker, J., "The importance of paradigms in the 21st century", presented at CF Business Forum, Bled (Slovenia), 20 March 2012. www.youtube.com/watch?v=6zyu 1yzEDHc.

13 AI and Big Data in Product Design

13.1 INTRODUCTION

During the last few decades the evolution of informatics has had a radical impact on society worldwide. The availability of powerful computers and of a network connecting them has allowed the creation of sophisticate software exploiting data availability in order to provide new services to users. The availability of usefulness of these services has generated in turn the rationale for a further increase in digitalization of information. This vortex of increasing and mutually feeding data processing power and data availability is well described by two keywords which are very much in the public consciousness today: "artificial intelligence" (AI) and "big data".

There are big expectations that these technologies will further improve the quality of human life and work in almost any field. It is thus a legitimate question now if they will apply to customer-centric product design too. Although the authors still have limited experience in systematically embedding these technologies into product design, it is worth closing this book trying to foresee how big data and AI could be used in customer-centric design in the near future.

13.2 AI AND BIG DATA: WHAT ARE THEY?

The growth in digital data availability is a consequence of increasing digitalization in society. However, technically speaking, "big data" [1] are distinguished from more traditional data types by their enormous size, the speed at which data are accumulated and processed, and their heterogeneity. Heterogeneity means that big data are not necessarily "structured" like traditional data types, that is, formed by a predefined sequence of fields which can be arranged in a table (a simple example of structured data includes employee name, date of birth, employment date, and job description in personnel records). On the contrary, big data can be either unstructured or a mix of these types. Many authors refer to these peculiar features as the 3 Vs of big data: volume (data size), velocity (of acquisition and processing), and variety (heterogeneity).

Examples of big data sources are social media, point-of-sales transactions and online banking, Global Positioning System (GPS) sensors, electronic health records, meteorological sensors, email, and documents shared within large organizations.

DOI: 10.1201/9781003544845-13

AI can be broadly defined as software that does the right thing, that is, it can solve the problem and achieve the best expected outcome [2] from the viewpoint of a human. This does not merely mean to follow an algorithm and provide an output based on the available input, which is what any software does. Instead, to be considered intelligent, a software must acquire and process an amount of data high enough to understand the context of the problem. Additionally, the algorithms used by AI are not limited to those defined by the programmers but can include those created by the software on its own by evaluating data, like in machine learning and deep learning technologies.

Consider, for instance, the problem of traveling from Paris to Rome by car in the minimum possible time. An "artificial stupidity" software would tell you that you can cover 1400 km by highway in the minimum possible time by constantly traveling at the highest possible speed offered by your car. If it is 180 km/h, you will have to press the gas pedal for about 8 hours. This may be a proper application of an algorithm, but such an answer would not be considered useful nor "intelligent" by most travelers. An "artificial intelligence" will consider the possible routes, speed limits and roadblocks along them, expected traffic conditions in the selected travel period, preferences of other travelers, etc., and suggest the best route options, the corresponding travel times, and any highway tolls. In this example you will probably get an estimate of about 14 hours, not to mention the time for necessary rest breaks along the way. This is what good navigation apps do today. Most humans judge such an answer as intelligent because neither they nor any of their friends could provide better advice.

Besides traveling support, there are many AI applications which anyone can already experience, and probably enjoy, in everyday life [2]:

- Generation of new content like text, images, video, audio, software code, etc., (in which case the term "generative intelligence" is typically used).
- Translation of text between different languages: online machine translation systems can cover the native languages of almost all humans with accuracy sufficient to understand a text. In specific cases, like two similar languages used by large populations, performances close to a human can be reached, especially if the text refers to a narrow domain of knowledge.
- Digital assistants which can answer questions and accomplish simple tasks like making calls, playing music, etc., like Amazon's Alexa and Apple's Siri.
- Provide recommendations: many of us already enjoy the accuracy of email spam filtering (which recommends not reading a mail), and words recommended by text editors while typing. AI-based recommendation systems are widely used by companies such as Amazon, Facebook, Netflix, Spotify, YouTube, etc., to recommend what you might also be interested in based on your past selections, or even the past selections of your contacts, with the final goal of promoting content by up-selling and cross-selling.
- Game playing: the supercomputer Deep Blue defeated the chess world champion Gary Kasparov already in 1997 and today AI-assisted game play ("Advanced Chess") is a widespread discipline on its own. You can train your chess or Go skills online against a machine, and you can set the difficulty level of the simulation software so high that no human has a chance to win.
- Facial recognition, which is today accurate enough to be used to secure mobile devices without having to type a password or to press a finger on the screen.

- Advanced domestic robots, such as automated vacuum cleaners and lawn mowers, rely on AI to avoid obstacles, do not fall from stairs, and identify the optimal path to follow.

Other important AI applications are not visible to the end-user, but are nevertheless very important, like fight against frauds (like credit card misuse and fake online reviews).

There are of course AI applications that do not make us happy all the time; typically, because they are not "intelligent" enough to fulfill our high expectations. An example is in customer service. AI can initiate a conversation with customers and direct them to the appropriate service agent, or even try to directly provide the service. If the AI simply retrieves information available as text elsewhere on the company website, many users will not hesitate to call it "stupid". In this case the AI provides the service quality of a cheap call center at an even cheaper price: it is convenient for the company, but frustrating for the user who would expect the "competent" advice of an expert knowing much more than what is publicly available.

Finally, a striking example of hoped-for application and big data to improve everybody's life is personalized medicine (also called precision medicine): today if we get sick, we must hope that our doctor has experience with our disease, formulates the correct diagnosis, and recommends the therapy fitting the disease and our body. These decisions are based on the experience and knowledge of the doctor, which is limited by what they can have learned and seen in their lifetime. The goal of precision medicine is to correlate the evidence of patient condition (not limited to the analyses evidencing the disease, but possibly including other information about patients such as their genetics) with medical data accumulated from millions or billions of other patients to identify the optimal therapy.

From the examples above we see in which cases AI can be most useful today:

1) Automation of tasks which could be performed by a human, but would be too expensive, boring, or would require a lot of training to learn the necessary skills. For instance, it is much easier to ask a digital assistant to dial a phone number while driving than to park the car, dial the number, activate the speaker, and start driving again (and it is much cheaper than hiring a secretary to travel with you). Or consider the time it would take to evaluate several possible routes on a map. Although automation is not a prerogative of AI, it is undoubtedly boosted by it.
2) Analysis of vast amounts of data, a task for which the human mind is not optimized.

From this discussion, the interdependency between AI and big data becomes clear too: AI needs a lot of data to understand the context and be useful; in the same fashion big data needs AI to be processed and to extract useful information from the rest.

13.3   APPLICATION OF AI IN INDUSTRY

AI offers to industry not only a potential product feature to delight customers but also as a tool to make businesses more effective and efficient [3], [4], [5].

The core idea behind the adoption of AI by industry is that ideally humans should focus on tasks which require "unique human" skills, while the rest could be delegated to machines. What these unique human skills are is a controversial philosophical question. Common answers are innovation, creativity, interpersonal skills, and empathy [6], although one could argue that these skills could be exhibited by a machine with sufficient processing power and training. The discussion of this dilemma, which would easily bring us to the meaning of the self and the existence of self-awareness, is clearly outside the scope of this book. What we want to focus on instead is the evidence that industry continuously seeks opportunity to improve its operations, and that AI has potential whenever the possibility arises that it can perform a task at the same level or better than humans, at the same cost or less.

Of course, the benefits of automation are not offered exclusively by AI which, however, could contribute considerably to it. McKinsey estimates that without generative AI automation could take over tasks accounting for about 22% of hours worked in the US by 2030, and that with generative AI this figure would rise to about 30% [7].

An interesting insight about the adoption of AI in industry is provided by another survey about the use of generative AI, which involved about 1400 participants, representing a broad range of regions, industries, company sizes, functional specialties, and tenures [8]. It turned out that about one company out of three already uses AI for activities that are covered in this book, the most common being marketing and sales (in particular content support for marketing strategy) and product development (particularly design and development, scientific literature and research review, and accelerated early simulation/testing). Note that software engineering, which can largely benefit from automated coding based on natural language specifications, was considered in this study as an independent voice separate from product development.

13.4 APPLICATION OF AI TO PRODUCT DESIGN

If we look more closely to the specific industry activity of product design, we can see two basic abilities of AI that are currently exploited [9], [10]:

Market and user research: AI tools can gather and make sense of market and consumer big data to assess the current situation (e.g. to uncover trends and patterns) and generate forecasts, with the final goal of supporting decisions about market needs and product features.

Concept development and testing: AI can quickly generate drawings and renditions of the product based on synthetic designer inputs, for instance, based on natural language. These visualizations can be used to better explain to customers and internal stakeholders what the product will look like in order to stimulate better feedback. In this sense, AI brings the prototyping capabilities already accelerated by CAD, 3D printing, and mockup software to a new level.

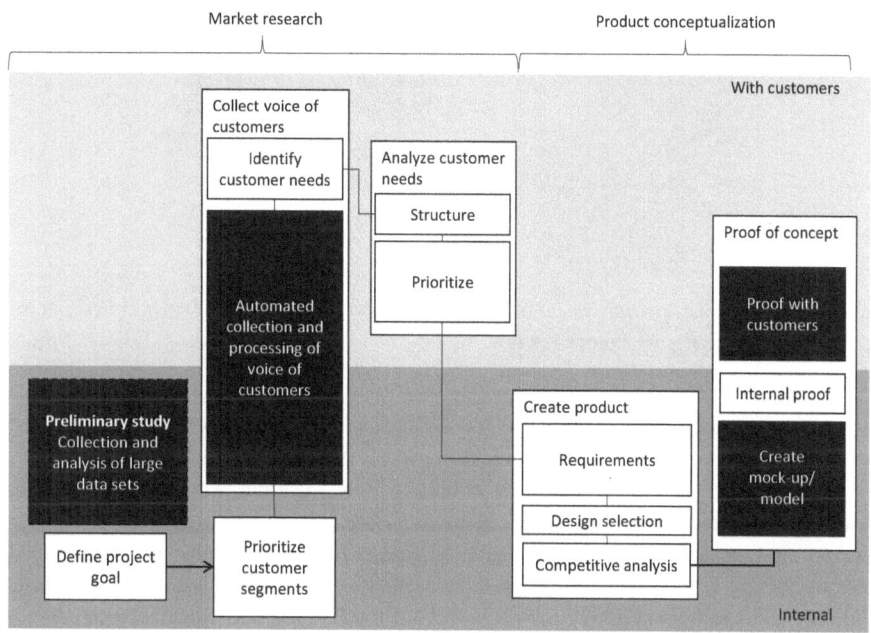

FIGURE 13.1   Stages in customer-centric product design which could benefit most from AI and big data are evidenced with black shading.

13.5   FUTURE PERSPECTIVES FOR CUSTOMER-CENTRIC PRODUCT DESIGN

If we compare the results of the previous section to the customer-centric product design workflow outlined in Figure 1.2, we can easily surmise which workflow stages could benefit most from AI and big data. This is shown in Figure 13.1, where these stages are highlighted with a black shading. Other steps require empathy between the designers and the customers, which is currently the prerogative of humans.

The step where the use of AI is the safest is the creation of prototypes, real-istic mockups, etc., which can be simplified and speeded up by generative AI. This represents an efficiency improvement, although not a game changer. We consider it a "safe application" of AI because results can be easily reviewed by designers, so that errors due to AI limitations can be identified and corrected.

The adoption of AI in other steps requires much more caution, since a review of results is difficult, or at least very time-consuming, and you will have to trust the results obtained by the machine. In all these cases a proper preliminary validation is necessary.

13.5.1   PRELIMINARY STUDY

For preliminary study you are probably already using AI to some extent – for instance, when you complement a systematic keyword-based search in selected publications

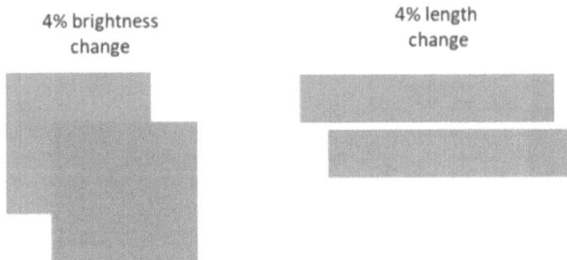

FIGURE 13.2 Relative comparison of geometrical shapes with tiny differences in brightness (left) and length (right). Referenced in Sec. 10.2.2.

with a broad search performed by a search engine like Google or Bing, which are likely to be powered by AI techniques. However, this kind of AI is used to find something you could have missed, and its accuracy is not critical. You may need to go further and use specific AI-based tools for the core of the preliminary study if you need to analyze a large amount of data. In this situation you can validate your AI tools by asking it to solve a problem whose answer you already know. An example could be performing a study you already did with great care in the past. It is relatively common today that professional software tools, marketed as AI-based, are offered to companies with the promise of gaining deeper insight into business problems. The first obvious validation step you should always start with is to ask about a problem in the field you are an expert in and see if the tool can reveal information that you were not aware of.

13.5.2 CUSTOMER FEEDBACK

Collection and processing of voices of customers can be automated if you want to deal with many customers. As we discussed in an earlier section, if you are interviewing customers face-to-face, you can obtain a good estimate of their needs with about 15 respondents from a uniform customer segment. The more you automate this task, the higher the risk of misunderstandings, which must be compensated for by increasing the size of the sample. The most obvious form of automation is to replace interviews with questionnaires, but you can go further by replacing the answers to explicit predefined questions with context-generated questions, or with the observation of customer behavior when using other products already on the market. The more the information you are looking for is hidden in large data sets, the more AI processing could help you to find it out.

Many products already in the market are constantly collecting data to monitor user behaviors as well as the health of the product in order to adapt the product operation to changing operation conditions. Such information can be used in product design to identify what goes right or wrong with the current product, and to identify new product features and customer benefits. These applications generate of course security concerns. Already, alarms have been raised in the QFD community on the use of grid square statistics to precisely track mobile phone users in real time. While the current purpose might be to facilitate emergency evacuations and other safety-related activities, the potential for abuse is being cautioned.

Similar considerations apply to the proof of concept. Presentation of the concept to customers and collection of their feedback can be automated in many ways. Again, the risk of misunderstanding must be compensated by feedback from many customers.

In general, it is possible that automated surveys of many customers not only provide the same accuracy of personalized interviews in less time but might even provide more reliable results. This relies on the quality of the automation process and on the intelligence of the AI.

13.5.3 Final Remarks

As a final remark, remember that good results are never free. We have warned already that the tools proposed do not constitute a magical recipe to solve all problems without effort. Similar considerations are valid for AI and big data. Big data are valuable not just because of their size, but because they may contain the information you are looking for. AI is valuable if, in view of its algorithms, training and input data, it has received the information necessary to solve your problem, and it has the processing capability to extract it. AI and big data are tools and, like any tool, the benefits they provide depend on their quality and the way they are used.

BIBLIOGRAPHY

[1] Balusamy, B., Abirami, N. R., Kadry, S., Gandomi, A. H., Big data: Concepts, technology, and architecture, Wiley, Hoboken (NJ), 2021.
[2] Norvig, P., Russel, S., Artificial intelligence: A modern approach, Global edition, Pearson, Harlow, 2021.
[3] Holdsworth, J., "The most valuable AI use cases for business", IBM Blog, February 14, 2024.
[4] Forbes. Q.ai Movement, "Applications of artificial intelligence across various industries", January 6, 2023. www.forbes.com/sites/qai/2023/01/06/applications-of-artificial-intelligence/
[5] Yüksel, N. et al. "Review of artificial intelligence applications in engineering design perspective", Engineering Applications of Artificial Intelligence, vol. 118, p. 105697, February 1, 2023.
[6] Huang, M.-H., Rust, R., Maksimovic, V., "The feeling economy: managing in the next generation of artificial intelligence (AI)", California Management Review, vol. 61, no. 4, pp. 43–65, 2019.
[7] Ellingrund, K. et al., Generative AI and the future of work in America, McKinsey Global Institute, McKinsey & Company, 2023.
[8] Singla, A. et al., The state of AI in early 2024: Gen AI adoption spikes and starts to generate value, McKinsey & Company, 2024.
[9] Booth, B., Donohew, J., Wu, W., Wlezien, C., How to use Generative AI in product design, McKinsey & Company, 2024.
[10] Cooper, R., "artificial intelligence and new product development – resources", Research-Technology Management, vol. 67, no. 1, pp. 70–75, January 1, 2024.

Appendix A
Why Not Averaging Global Priorities?

Throughout this book, global priorities of customer needs have been calculated using the averaged local priorities. You could have equally followed a different procedure by (1) calculating first the global priorities for every customer and (2) averaging them. As we have already mentioned, these two methods typically provide equivalent results, and significant discrepancies can arise only if you have too few customers, or if the customer segment is not homogeneous. This fact can be better understood with a simple example.

Consider two needs a and b in the hierarchy diagram, where a is the parent of b. Let P_a be the global priority of need a and P_b the local priority of need b. The global priority of b is $P_a \cdot P_b$. Its expected value $E(P_a \cdot P_b)$ is related to the expected values of P_a and P_b by the formula:

$$E(P_a \cdot P_b) = E(P_a) \cdot E(P_b) + Cov(P_a, P_b). \tag{A.1}$$

This formula is a direct consequence of the definition of the quantity $Cov(P_a, P_b)$, which is called "covariance" and indicates how the fluctuations P_a and P_b about their mean values are correlated:

$$Cov(P_a, P_b) = E\big((P_a - E(P_a)) \cdot (P_b - E(P_b))\big). \tag{A.2}$$

Expected values of random variables are approximated by their average over many customers (let us say N customers), for instance:

$$E(P_a) \approx \bar{P}_a = \frac{1}{N}(P_{a1} + \cdots + P_{aN}). \tag{A.3}$$

$$E(P_a \cdot P_b) \approx \overline{P_a \cdot P_b} = \frac{1}{N}(P_{a1} \cdot P_{b1} + \cdots + P_{aN} \cdot P_{bN}). \tag{A.4}$$

This is explained in the example of Figure A.1. If an increase of P_a typically corresponds to an increase in P_b, then the covariance is positive (Figure A.1, left). If

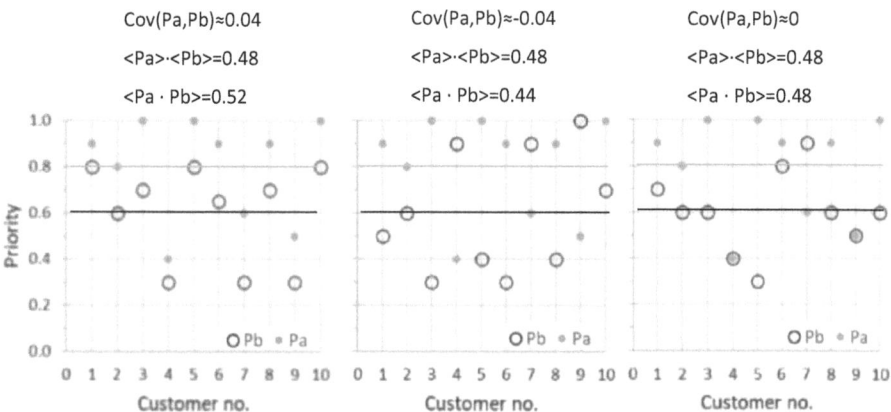

FIGURE A.1 Three sets of customer judgments for the global priority of a parent need (P_a) and the local priority of its child need (P_b). Left: judgments with positive covariance; middle: judgments with negative covariance; right: independent judgments. Solid lines indicate the average values of P_a (gray) and P_b (black).

an increase in P_a corresponds to a decrease in P_b, then the covariance is negative (Figure A.1 middle). If the changes in P_a are independent of the changes in P_b, then the covariance is 0 and the global priority is independent of the way it is calculated (Figure A.1, right).

In our case the local priority of a need within the group is independent of the priority of its parent: the fact that one need is preferred over a sibling in the group does not depend on who the parent is. Thus, we can expect that the covariance is null, and the average global priority is equal to the global priority calculated with average local priorities.

Appendix B
Equivalence of Different
AHP Scales

We have already mentioned that although the 1–9 scale and the geometric progression are significantly different, they substantially generate the same priorities when used in AHP. This can be easily seen with an example with five elements E1–E5 such that:

- E1 is moderately more important than E2, strongly more important than E3, very strongly more important than E4, and extremely more important than E5.
- The judgments are substantially consistent, for example, E2 is moderately more important than E3, etc.

With the 1–9 scale, the judgment matrix is:

$$
\begin{pmatrix}
1 & 3 & 5 & 7 & 9 \\
1/3 & 1 & 3 & 5 & 7 \\
1/5 & 1/3 & 1 & 3 & 5 \\
1/7 & 1/5 & 1/3 & 1 & 3 \\
1/9 & 1/7 & 1/5 & 1/3 & 1
\end{pmatrix}.
$$

It the geometric progression is used instead, the judgment matrix is:

$$
\begin{pmatrix}
1 & 2 & 4 & 8 & 16 \\
1/2 & 1 & 2 & 4 & 8 \\
1/4 & 1/2 & 1 & 2 & 4 \\
1/8 & 1/4 & 1/2 & 1 & 2 \\
1/16 & 1/8 & 1/4 & 1/2 & 1
\end{pmatrix}.
$$

TABLE B.1
Comparison of results obtained with two different scales

1–9 Scale			2^n Scale			
Priorities P_1–P_5	λ	CR	Priorities P_1–P_5	λ	CR	RMS Priority Difference
0.513			0.516			
0.261			0.258			
0.129	5.24	0.053	0.129	5	0	0.002
0.063			0.065			
0.033			0.032			

Note: λ indicates the leading eigenvalue of the judgment matrix and CR the consistency ratio.

The results are shown in Table B.1. With the 2^n scale the matrix is perfectly consistent (CR = 0, λ = 5) and priorities follow exactly a geometric progression. With the "1–9 scale" priorities approximate well the geometric progression (the RMS error is just 0.002) and the consistency ratio is below our tolerance threshold (0.05<0.1).

Appendix C
Relationship with ISO 16355

Most subjects covered in this book are also present in the standard ISO 16355: "Applications of statistical and related methods to new technology and product development process". You can find in Table C.1 the parts and sections of the ISO standard where different subjects are discussed.

TABLE C.1
Correspondence between subjects discussed in this book and sections in the standard ISO 16355

Subject in this book	Part	Section
		Reference in ISO 16355
Project goal	1	A2 Project goals table
	2	9.1.3. Business goals for projects
Customer segmentation	1	A9 Customer segments table
	2	9.2.2.2. Customer segments table
Customer value chain	2	9.2.1.2. Customer value chain
Customer contact checklist	2	9.2.5.2.2 Gemba visit checklist
Customer process model	1	A.8 Annotated customer process model
	2	9.2.5.2.3. Customer process model
Customer visit/interview table	1	A.9 Gemba visit table
	2	9.2.5.2.4. Gemba visit table
Customer voice table	4	9.2.4 Customer voice table
	1	A.10 Customer voice table
Affinity diagram	4	10.2 Affinity diagram
	1	A.11 Affinity diagram for customer needs
Hierarchy diagram	4	10.3 Hierarchy diagram
	1	A.12 Hierarchy diagram of customer needs
Pairwise comparison	4	11.2 Applying AHP to customer needs
	4	11.3 Steps to AHP using a spreadsheet
	4	11.6 Applying AHP to a customer needs hierarchy
	1	A.13 Customer needs prioritization with analytic hierarchy process

TABLE C.1 (Continued)
Correspondence between subjects discussed in this book and sections in the standard ISO 16355

Subject in this book	Reference in ISO 16355	
	Part	Section
Tables	1	A.15 Maximum value table
	5	9.2 Maximum value table
Qualitative matrices	5	9.3 L-matrices
Quantitative matrices	1	A.16 Customer needs/functional requirements matrix (House of Quality) and other L-Matrices
	5	10.2 Transfer of prioritization
Design selection	1	A.17 Design planning table
	5	10.3.4.1 Design planning table for the house of quality
Competitive analysis	1	A.14 Quality planning table (unweighted)
	4	12.2 Quality planning table
Kano method	5	10.3.4.4 Traditional and new Kano model
Basic principles	1	4.1 The theory and principles of QFD
Super Pugh method	1	A.18 Super Pugh concept selection with AHP
	5	10.4.3.7.2. Selecting concepts using Pugh and Super Pugh methods
Beyond product design	5	10.4 Transferring deployment sets by dimensions and levels
	8	(The entire volume is dedicated to quality deployment through the full product lifecycle)

Index

Note: Figures are indicated by *italics*. Tables are indicated by **bold**. Endnotes are indicated by the page number followed by 'n' and the endnote number e.g., 20n1 refers to endnote 1 on page 20.